Symbiogenesis
A Macro-Mechanism of Evolution

Cover illustration
Hypothetical overall picture of evolution from big bang to man, depicted as an evolutionary spiral. On the basis of the pulsation theory it can be supposed that after a collapse of matter in the entire cosmos the process of the cosmic, chemical, biological and cultural phases of evolution may recommence, initiated by another big bang.

Werner Schwemmler

Symbiogenesis
A Macro-Mechanism of Evolution

Progress Towards a Unified Theory of Evolution Based on Studies in Cell Biology

W DE G

Walter de Gruyter · Berlin · New York 1989

Prof. Dr. Werner Schwemmler
Freie Universität Berlin
Institut für Pflanzenphysiologie, Zellbiologie und Mikrobiologie
Königin-Luise-Straße 12–16 a
D-1000 Berlin 33
Federal Republic of Germany

Library of Congress Cataloging in Publication Data

Schwemmler, Werner, 1940–
Symbiogenesis : a macro-mechanism of evolution : progress towards a unified theory of evolution based on studies in cell biology / Werner Schwemmler.
p. cm.
Bibliography: p.
Includes index.
ISBN 0-89925-589-2 (U.S.)
1. Evolution. 2. Life--Origin. 3. Cosmology. I. Title.
QH371.S393 1989
575.01--dc20 89-7804

Deutsche Bibliothek Cataloging in Publication Data

Schwemmler, Werner:
Symbiogenesis, a macro-mechanism of evolution : progress towards a unified theory of evolution based on studies in cell biology / Werner Schwemmler. – Berlin ; New York : de Gruyter, 1989
ISBN 3-11-012132-8

Printing: Gerike GmbH, Berlin. – Binding: Lüderitz & Bauer GmbH, Berlin. – Printed in Germany.

Foreword

Colleagues who have read parts of the manuscript for this book have suggested that an overview of the whole would make it much easier for them to judge the merit of individual chapters. In response to this helpful criticism, I should like to present briefly my reasons for writing a book of this nature, before going into the individual chapters.

This book aims at providing the scientifically-interested reader with an overview of the entire complex of evolution, beginning with the origins of the universe and continuing through the emergence of the human life form. The sweep of evolution is so broad and so complex that it cannot be understood without being broken down into smaller, more manageable, parts. What could be more logical than to proceed chronologically, dividing evolution into cosmic, chemical, biological and cultural phases? These phases are marked by the formation of elementary particles, atoms, molecules, cells and multicellular organisms, including human beings, the "carriers" of culture. The description of the structures and functions of these basic units seems to be much easier to understand than the tracing of the exact pathway of their development.

In the first chapter, the goals and the methodological difficulties of reconstructing the entire evolutionary process are considered. In the second and third chapters, current hypotheses on the course of the cosmic and chemical phases of evolution are presented and compared to available data, in order to determine which appears most likely to be correct. Then, on the basis of the most likely hypotheses, those phases of evolution are reconstructed. This reconstruction does not pretend to be complete and may well be a selective, subjective interpretation of the available data. The reader is entirely free to draw other conclusions from the data. In any case, by grappling with the problems of reconstruction presented, the reader should come to a deeper understanding of biogenesis and its mechanisms. No effort is made to present a comprehensive collection of facts relating to cosmo- and chemogenesis, but only to select those which provide a better understanding of biogenesis.

The fourth chapter treats the actual processes of biogenesis. Important speculations on possible transitions from molecules to cells, on the rather uncharted evolutionary "leap" to bacterial cells, and finally, on the evolution of higher (eukaryotic) cells, are discussed here. The treatment of the material is the same in all three subchapters: first a brief presentation of the evolutionary systems in question, then the most important alternative hypotheses and their origins. This is followed by a presentation of the facts pertinent to any reconstruction of the evolutionary process, and then by our own attempt at reconstruction. Here again, the facts selected for presentation represent only the essential aspects of evolution. Comprehensive collections of facts can be found in the numerous text books and monographs listed at the end of each chapter.

Chapter Five deals with the phylogenetic roots of human sociogenesis which preceded the latest phase of evolution, cultural evolution. The treatment of human beings as biological and cultural beings follows the same methodological pattern as was maintained for other elements in the respective discussions of the preceding phases.

Chapter Six deals with the widely held opinion that selection and mutation alone are not sufficient to achieve the macro-evolutionary transitions from one basic biological type to the next. Mechanisms are sketched here which may help explain those changes not adequately explained by the micro-evolutionary mechanisms. The formation of symbioses (symbiogenesis) is emphasized as one important macro-evolutionary mechanism.

Finally, in the last chapter, an experimental system is presented which offers the possibility of testing some of the hypotheses discussed earlier in the book. The particular insect system examined may provide evidence that the major consequence of symbiogenesis is the periodicity in all evolutionary systems which can be applied to unsolved problems of cell biology, such as cell differentiation, de-differentiation and cell rhythm. Periodicity could also provide a theoretical basis for biology which is directly linked to the established theories of physics and chemistry. Classification of cells into a preliminary periodic system represents the logical consequence to the particular theoretical background which has been selected.

I wish to express my appreciation to the following colleagues who prolificly discussed the manuscript or gave criticism: C.G.Arnold, C.F.Bardele, T.Cavalier-Smith, F.Crosby, K.Dose, B.C.Goodwin, W.Kaplan, G.Kraepelin, C.Ponnamperuma, H.Schenk, U.Schönrock, A.Schwartz, H.Schwemmler, P.Sitte, E.Stackebrandt, J.Wolters, A.Zimmermann.

I am most grateful to Mary Eagleson for translating the manuscript into English and for giving criticism of great value.

My thanks also go to D.Rössing and H.Kulike for their intensive assessments and helpful suggestions.

Finally I must pay tribute to the friendly cooperation and patience with which the publisher, Walter de Gruyter, encouraged my work.

Berlin 1989 Werner Schwemmler

Table of Contents

Chapter 4 Biogenesis

First subchapter Precyte

Second subchapter Procyte

Third subchapter Eucyte

Chapter 5: Sociogenesis

Chapter 6: Macro-mechanisms of Evolution

Chapter 7: Experimental application

Appendix

Chapter 1 Evolution Research

Space travel has opened up new dimensions for humanity which has begun to explore neighbouring cosmic systems, and which may even settle in them at some future time. Thus it is of great importance before leaving our planet, to briefly take stock and summarize the essentials of scientific knowledge concerning the origin and development of our species.

A science usually first describes its phenomena then analyzes them by searching for regularities and finally, arranges them in a classification scheme. In this respect, physics and chemistry are classical scientific disciplines. They have described and analyzed the basic phenomena in their fields of inquiry, and have also classified them in systems of elementary particles and in the periodic system of the elements. A prerequisite for this classification was an understanding of structure and properties of the smallest units of matter. The periodic system of the elements was conceived on the basis of Dalton's atomic hypothesis, and was later amply confirmed and elucidated by the quantum theory.

In biology, basic phenomena and conformities of unicellular and multicellular organisms have been described, analyzed and classified (evolution research, genetics, molecular biology...). However, the smallest living components of biological systems, the cells, have not yet been arranged in a system of classification. Nevertheless, the theory of evolution and the theory of the origin of cells, confirmed by experiments, provide sufficient explanations concerning structure, function, information content and evolution of cells. Thus it should now be possible for biologists to set up a system in which to arrange the smallest cellular components, and thus to contribute to theoretical biology. Perhaps such an undertaking will also provide new experimental impetus for biology.

Basic theories of biology, e.g. **the theory of evolution** and the related **endocytobiotic cell theory**, are presented in the following in an outline form, using cell evolution as an example. On this basis, a **periodic system of cells** is then proposed. However, before we can turn to this task a few words must be said about the underlying theory of knowledge (epistemology) and about the methodological problems of such an attempt.

I. The epistemological problem

The **scientist** thinks logically about the internal and external phenomena of the real world[1] as revealed by **observation** and **experiment** (Fig.1). The results are objective scientific data which, under the same conditions, can be reproduced at any time, anywhere. The objects of science, such as atoms, molecules, cells, and fungus, plant and animal organisms are less complex than the cognitive human apparatus analyzing them.

Reality can also be **experienced** in a primarily intuitive way. This leads to the data of the **arts**, reproducible only statistically within certain limits. In this case the objects, e.g. the actions, thoughts, feelings and the speech of human beings, and thus sometimes the human sensory apparatus are by nature of the same degree of complexity as the cognitive apparatus which is examining them.

Finally, **metaphysics** attempts to say something about the world as a whole and the position of human beings in it. These phenomena are many times more complex than the human cognitive and sensory apparatus exploring them. Metaphysical phenomena are studied, for example, in psychology, philosophy and theology. We comprehend such statements at best **synoptically** - as a type of intuitive and emotional reflection of the real world as a whole.

II. The methodological problems

Evolution - in the broadest sense - is the process leading from the origin of the universe to the formation of our world with its living beings, including humans with their cultural awareness. This process can be divided into **physical** (cosmic), **chemical, biological and cultural phases.** The essential milestones passed in the course of this process are the atom, the molecule, the cell and the human

---

[1] The 'real world' is defined as the sum of phenomena in space and time we are able to perceive within ourselves and in our surroundings.

Fig. 1. Correlation between the degree of complexity of the human cognitive apparatus, methods of examination, and statements about the real world (for details, see text above).

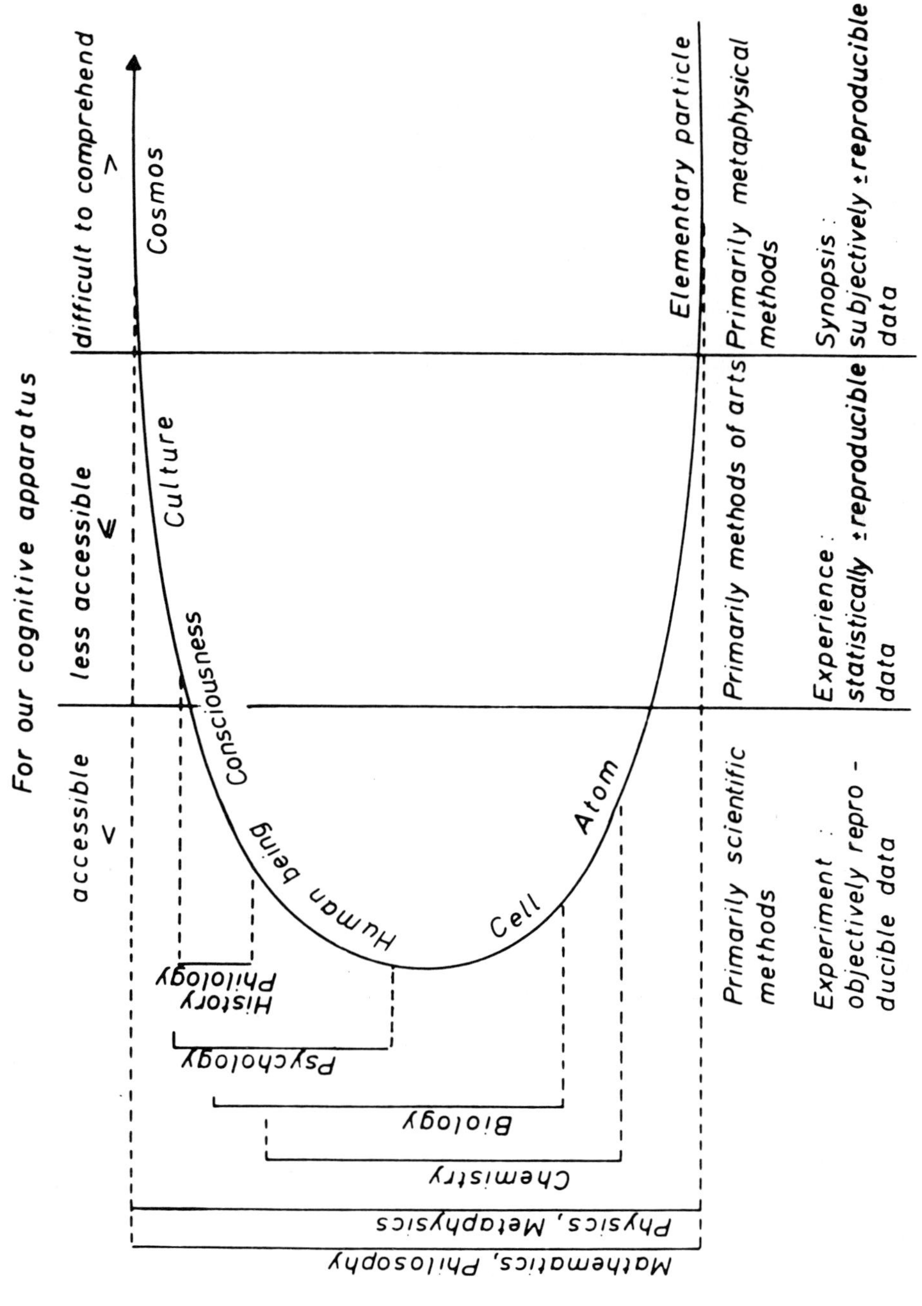
For our cognitive apparatus
accessible
<
less accessible
≤
difficult to comprehend
>
Cosmos
Culture
Consciousness
Human being
Cell
Atom
Elementary particle
History
Philology
Psychology
Biology
Chemistry
Physics, Metaphysics
Mathematics, Philosophy
Primarily scientific methods
Primarily methods of arts
Primarily metaphysical methods
Experiment : objectively repro - ducible data
Experience : statistically ± reproducible data
Synopsis : subjectively ± reproducible data

individual, as a bio-cultural being. Thus evolution is the most complex, and at the same time, the central phenomenon of biology, indeed of the real world as a whole. **Evolution can only be thoroughly examined through the cooperative efforts of all three branches of knowledge: sciences, arts and metaphysics.** The arts and metaphysics deal more with the individual aspects of the evolutionary system being characterized by its degree of freedom, while the sciences concentrate on the causal, material or energetic aspects of the evolutionary system.

A further problem is that the evolutionary process, which has been taking place for billions upon billions of years, is a natural historical phenomenon which cannot be reversed, and which we can now only hope to reconstruct from direct and indirect evidence, e.g. fossils and homologies. The evidence is obtained by observation, experience and/or experiment, and provides the basis for **working hypotheses**, the implications of which extend beyond the individual observation or experience. These working hypotheses again are the basis for predictions which may be tested experimentally. A hypothesis becomes a **rule** or **law** if it has been confirmed by all experimental tests, and has not been contradicted by any. A number of rules or laws which can be explained by common principles are summarized as **theories**. The next step is the attempt to reconcile several different theories with a larger concept. If in this way one arrives at a concept combining several partial theories, one speaks of a **complex theory**, such as the **theory of evolution**. The statements of evolution research are thus always a mixture, on the one hand, of observations, experiences and experiments, and on the other, of hypotheses, rules, laws and theories, which in their various forms are the results of thoughts. The process of obtaining knowledge simply by evaluation of observation is called induction, while the drawing of conclusions through application of logic to the available data is called deduction. Both types of reasoning are used in science.

The theory of cosmogenesis is thus a derived theory on the origin and development of the entire cosmos. The theory of evolution may be broken down into subtheories: biogenesis, the development of life from non-living precursors, and chemogenesis, the formation of the molecular precursors of life from much simpler forms of matter early in the earth's history.

Scientific evolution research thus always involves deduction of a chain of events from evolution theories. A methodologically reasonable way is first to present the alternative working hypotheses applicable to a phase of evolution, then to examine the available data to determine whether they fit better one or the other hypotheses (in the sense that what confirms one hypothesis may contradict the other), and finally to take the more probable hypothesis as the basis for reconstructing the evolutionary process in question. A combination of verbal and graphic means seems to be the best method for turning apparently parallel aspects of the complex process of evolution into a simple linear concept.

III. References

Ayala, F.J. (1979): *The Mechanisms of Evolution*.Scientific American 9 48-61.

Bresch, C. (1977): *Zwischenstufe Leben; Evolution ohne Ziel?* Piper Verlag München, Zürich.

Capra, F. (1979): *The Tao of Physics. An exploration of the parallels between modern physics and eastern mysticism*. Wildwood House London, (1975, Repr.).

Ditfurth, H. von (1981): *Wir sind nicht nur von dieser Welt; Religion und die Zukunft des Menschen*. Hoffmann und Campe, Hamburg.

Kimura, M. (1979): *The Neutral Theory of Molecular Evolution*. Scientific American 11 94-104.

Osche, G. (1974): *Evolution: Grundlagen, Erkenntnisse, Entwicklungen der Abstammungslehre*. 3rd ed., Herder Verlag Freiburg, Basel, Wien.

Peters, D.S., Franzen, J.L., Gutmann, W.F., Mollenhauer, D. (1974): *Evolutionstheorie und Rekonstruktion des stammesgeschichtlichen Ablaufs*. Umschau 74 501-506.

Prigogine, I., Stengers, I. (1981): *Dialog mit der Natur. Neue Wege naturwissenschaftlichen Denkens*. 2nd ed., Piper Verlag München.

Riedl, R. (1980): *Biologie der Erkenntnis. Die stammesgeschichtlichen Grundlagen der Vernunft*. 2nd ed., Paul Parey Verlag Hamburg, Berlin.

Schwemmler, W. (1982): *The endocytobiotic cell theory and the periodic system of cells*. Acta Biotheoretica 31 45-68.

\- (1984): *Reconstruction of cell evolution: A periodic system*. CRC Press, Boca Raton, Florida.

Chapter 2 Cosmogenesis

I. Phenomenon

In our world - that part of the cosmos which we can view with technical aid (radio telescopes, reflector-type telescopes) - there are at least six hierarchic cosmic systems which each fit into the next largest. In decreasing order of complexity, these are the **galactic cluster**, **galaxy** (spiral nebula), **globular star**, **star**, **planet** and **satellite** (Fig. 2). Since the light of any celestial body requires a certain amount of time to reach our telescopes - the sensitivity of telescopes is at present limited to 10^{10} light years - we see each system at the stage of development it had reached when the light left it; in the case of the most distant systems, this was billions of years ago. Based partly on this observation of the past, and its evaluation by means of telescopy, spectral analysis, measurements of gravitational and magnetic fields, geometry, kinematics and particle physics, and assisted by experiments carried out by space probes, it is now possible to develop experimentally-based hypotheses about the origin and development of the cosmos (e.g. Fritzsch 1983).

The term **cosmos includes space, time and matter.** The cosmos is therefore characterized by at least four coordinates: three of space or matter, and one of time.

II. Hypotheses

The broad spectrum of ideas discussed in the past range from the **steady state** to the opposing **big bang hypothesis**. The **steady state hypothesis**, going back to the germinal ideas of **Newton** (1643-1727), postulates that the material distribution of the universe is homogeneous and very stable; that the universe is infinite and space is static. These assumptions are neither consistent with the fact that the measured forces of gravity are always finite nor with the fact that forces have been described which effect only short distances. **Einstein** attempted to correct these shortcomings in the Newtonian concept with his **theory of gravitation** (1905). He assumed a universe which developed from nothing, which can be represented either as a "negatively curved", infinitely expanding open space or as a "positively curved" and thus closed or finite space (Fig. 3). The **pulsation hypothesis**, suggested by **Hughes** (1963), indicates one condition for a finite universe. According to this hypothesis, the cosmos began with an enormous explosion, the so-called **big bang**, and will eventually reach a maximum expansion, after which it will collapse upon itself, leading to another big bang and thus initiating the next cosmic cycle. More recent modifying hypotheses, such as the saddle or bubble model, will not be investigated in this brief survey.

III. Data

Essential experimental data for the derivation of cosmogenesis is examined in the following to determine whether they support one or the other alternative hypotheses. The more plausible hypothesis will be applied to reconstruct one possibility of cosmogenesis.

A. Redshift

In 1923, **Hubble** discovered that the light reaching the earth from distant galaxies shifts toward the red end of the electromagnetic spectrum. It can be deduced from this shift that all these galaxies are moving away from the earth, i.e. that the entire cosmic space is expanding. From the same notion, one can extrapolate back to the time at which the expansion started. This procedure is logically

---

Fig. 2. Hierarchy and developmental phases of the cosmic systems in the observable portion of the universe, from the galactic clusters to galaxies, globular stars, stars, planets, and their satellites. This presentation of the universe should not be taken as a three-dimensional model; the time dimension at least should be taken into account as one further dimension. The schemes are not drawn to scale with respect to each other (arrangement of data by Schwemmler, 1984).

* 1 billion = 10^9

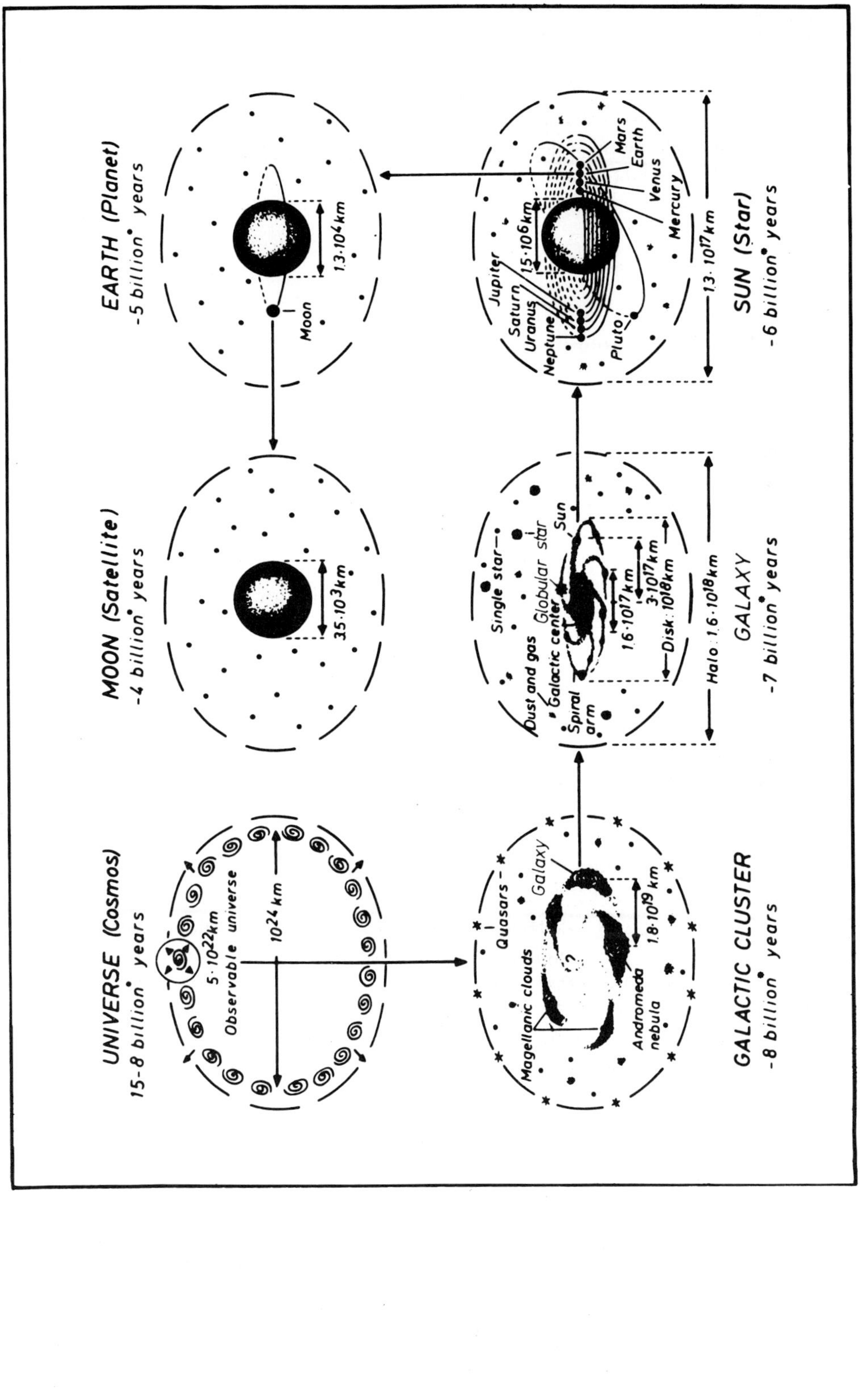
UNIVERSE (Cosmos)
15-8 billion years
5·10²²km
Observable universe
10²⁴ km
MOON (Satellite)
-4 billion years
35·10³km
EARTH (Planet)
-5 billion years
1.3·10⁴km
Moon
GALACTIC CLUSTER
-8 billion years
Magellanic clouds
Quasars
Galaxy
Andromeda nebula
1.8·10¹⁹ km
GALAXY
-7 billion years
Dust and gas
Galactic center
Spiral arm
Single star
Globular star
Sun
1.6·10¹⁷km
3·10¹⁷km
Disk 10¹⁸km
Halo 1.6·10¹⁸km
SUN (Star)
-6 billion years
Jupiter
Saturn
Uranus
Neptune
Pluto
15·10⁶km
Mars
Earth
Venus
Mercury
1.3·10¹⁷km

justified because **there is a linear correlation between the distance of a galaxy from the observer and its relative velocity**. For every million light years[1] distance from us, the velocity of the expansion increases by about 17 km/sec. Since the velocity of the light is constant, the distance is a linear function of the time which has passed; therefore there is also a **linear relationship between the expansion velocity and the age of a galaxy**: the further away a galaxy is from the observer, and the longer it has been expanding, the greater is the velocity of the expansion. These data can be interpreted by a initial explosion, or big bang, as this original "singularity" is called.

B. Background radiation

A second discovery which had already been predicted by cosmologists also supports the big bang hypothesis. In 1964, **Penzias** and **Wilson** found a radiation in the microwave range with a wavelength of 7.35 cm which arrives from all directions of the cosmos with practically the same intensity; it is isotropic. The radiation has a temperature of about 3^0 Kelvin (-270 ^{0}C). **From the presently very weak glow of the radiation, it can be extrapolated that 10^{10} - 2×10^{10} years ago the cosmos had a very high temperature.** In addition, at that time the density

[1] A light year is the distance light travels in one year: at a velocity of about 300,000 km/sec, it moves about 9.5×10^{12} km/year.

Fig. 3. Relativistic model of a **closed, finite universe** with an age of about 8×10^9 years (**Friedmann** time): t signifies time, t_0 the present, R(t) the radius as a function of time, which describes the expansion of the universe, R_0 is the present radius of the universe, q the acceleration of expansion of the celestial bodies, and q_0 their present acceleration. The tangent to the R(t) curve represents the model of an **open, infinite universe**, with an age of 16×10^9 years (**Hubble** time). It is obtained by linear extrapolation of the expansion equation, according to which the velocity of the celestial bodies increases in proportion to their distance. Neither model can be proven, since the analysis of the biggest as well as the smallest system (in physics) must rely on basic assumptions which are - as yet - unproven (adapted from Dehnen 1976).

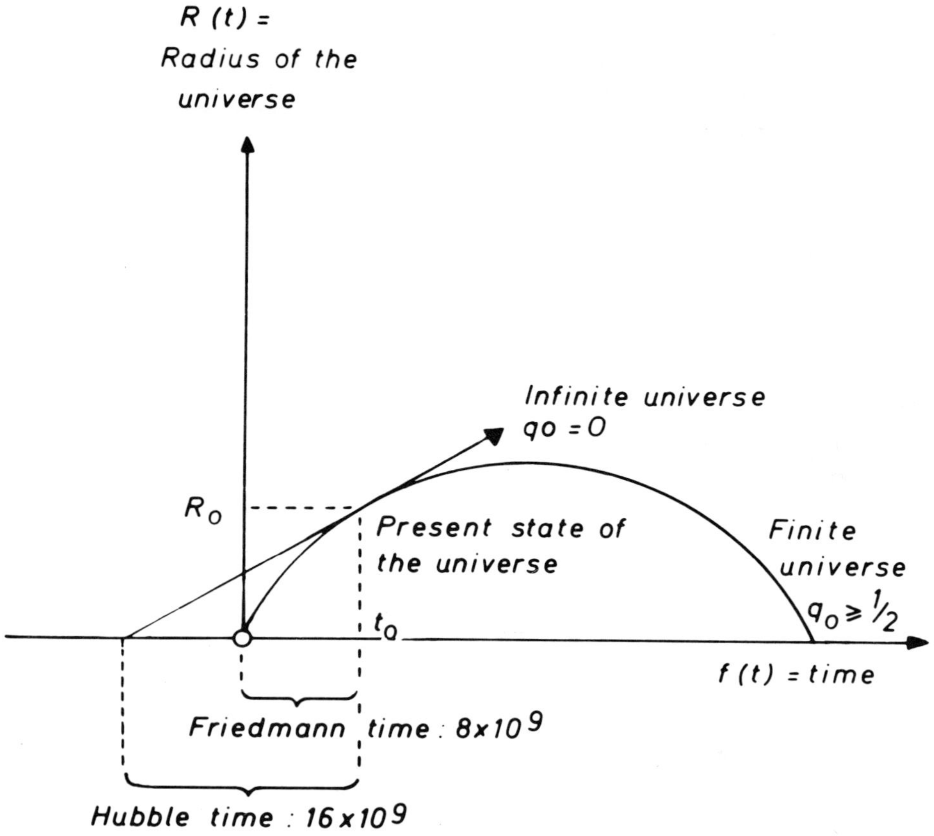
R (t) =
Radius of the
universe
Infinite universe
qo = 0
R_0
Present state of
the universe
Finite
universe
$q_0 \geq 1/2$
t_0
f (t) = time
Friedmann time : 8x10⁹
Hubble time : 16x10⁹

of matter must have been so great that atoms were not stable, but would have decayed into elementary particles.

Temperatures and densities as great as those which must have existed at the time of the extrapolated big bang cannot be reproduced in the laboratory, so that direct experimental verification of the derived data is not possible. However, elementary particle physics provides many new insights into the structures and properties of the smallest components of matter under extreme conditions, and these in turn provide support for the cosmological considerations. The partially successful efforts to create a unified theory of the basic physical forces (strong interaction, weak interaction, electromagnetism, gravitation) have also provided further confirmatory evidence.

C. Critical density

From measurements of the amount of matter in a sufficiently large volume of space, it is possible to estimate the average density of the entire universe. The data presently available suggest that this density is only slightly less than the so-called "critical density" of 10^{-29}g/cm^3 (= 10 hydrogen atoms per m^3). **Critical density is attained when the effect of gravity exceeds the momentum of expansion, so that the universe will contract.** Cosmologists are currently investigating whether matter in the universe exhibits critical density. If so, the universe is closed, but if not, the universe is open. In the latter case, the expansion would be slowed down by gravitation, but not completely brought to a halt. An intermediate stage between closed and open is not possible. Estimates based on measurements of the light of stars and other luminous matter suggest that the total mass is about 10% of the critical mass. However, such estimates do not adequately account for the gravitational motions of galaxies, and it is thought that they must contain large amounts of dark matter (cp. black holes[1]) in order to move as they do. The dark matter might make up a considerable fraction of the total mass of the universe, possibly enough to bring it to the critical density and to close it.

---

[1] A **black hole** is accumulated matter extremely condensed at the expense of E_{kin} and is the last stage of "superheavy" stars following a gravitational collapse. Neither particles nor radiation can escape from its force of attraction; in a sense, a black hole has closed itself off from the rest of the universe.

D. Particle interaction

According to calculations of elementary particle interactions made by Dicus et al.(1983), the universe could be cyclic. It would then go through alternating phases of expansion and contraction, assuming that the density of matter is great enough to close it; after the gravitational collapse, another big bang could take place. In this case, the expansion phase would become longer with each cycle, if the possible energy gains from each contraction were conserved until the next expansion. **Dicus et al.** paid special attention to the changes of energy in a **pulsating universe**, because not only do the distance between two points in space change during cosmic expansion and contraction, but also the wavelength and therefore the energy of photons. If the universe behaves in a manner analogous to the surface of a balloon[1] during expansion, the wavelength of the photons will increase. Since the energy of a photon is inversely proportional to its wavelength, it would decrease during expansion and increase during contraction. A photon emitted toward the end of expansion can thus gain energy, since its wavelength would become shorter during contraction than it was at the time of emission. Energy gained by the universe in this way could cause the next expansion phase to be decreased, i.e. according to the formula $E = m \cdot c^2/\lambda$. The formula equates the increase of cosmic energy with the increase of matter in the following cycle, which means a decrease in length of the cycle. On the other hand a photon emitted at the beginning of expansion would not lose energy.

IV. Reconstruction

A. The problem

Evidence unequivocally supports the big bang hypothesis, i.e. an evolving universe, rather than a stable and static universe. It also appears possible that comprehensive studies of the total cosmic mass, including all dark matter (such as black holes), will detect a value corresponding to the critical density required

---

[1] but as a three-dimensional system in a four-dimensional space.

to close the universe. A number of considerations - as mentioned above - suggest that the universe is cyclic, oscillating continuously between collapse and big bang. However, far too little is known about the mechanisms by which such a pulsing universe could keep going. There is also no answer to the question whether the extrapolated cosmic cycles would last for the same or a different length of time.

A number of other decisive problems have not yet been settled (Zimmermann 1985):

> *1) The universe is, as far as we can observe, homogeneous on a large scale. For this* **homogeneity**, *the big bang hypothesis offers no explanation. Some cosmologists assume, therefore, that it was a highly improbable type of initial singularity. There is also no consensus on the type of symmetry which determines the unified forces.*
>
> *2) The large-scale homogeneity of matter distribution is accompanied by very great* **inhomogeneities** *on smaller scales, such as the galaxies, galactic clusters and other enormous collections of material structures. There is still no explanation for this distribution of matter.*
>
> *3) The fact that our cosmos is probably constructed (for the most part) of* **matter**, *and not* **antimatter** *(which, to the best of our knowledge, must exist) still lacks explanation.*

On the whole, however, it appears that the explanation of the origin and development of the universe offered by the pulsation hypothesis agrees best with the available data (Davies 1984), so that the reconstruction of cosmogenesis presented here in outline form will be based on this hypothesis .

Fig. 4. Hypothetical scheme of a **developmental cycle for the entire universe**, including the formation of light and heavy atoms, based on the pulsation theory. The representation of the universe is not intended to be a three-dimensional model. It is based rather on the concept of a three-dimensional balloon expanding in a four-dimensional space, the fourth dimension being time (arrangement of data by Schwemmler 1984, for details see Dehnen 1976 and Davies 1984).

* 1 billion = 10^9

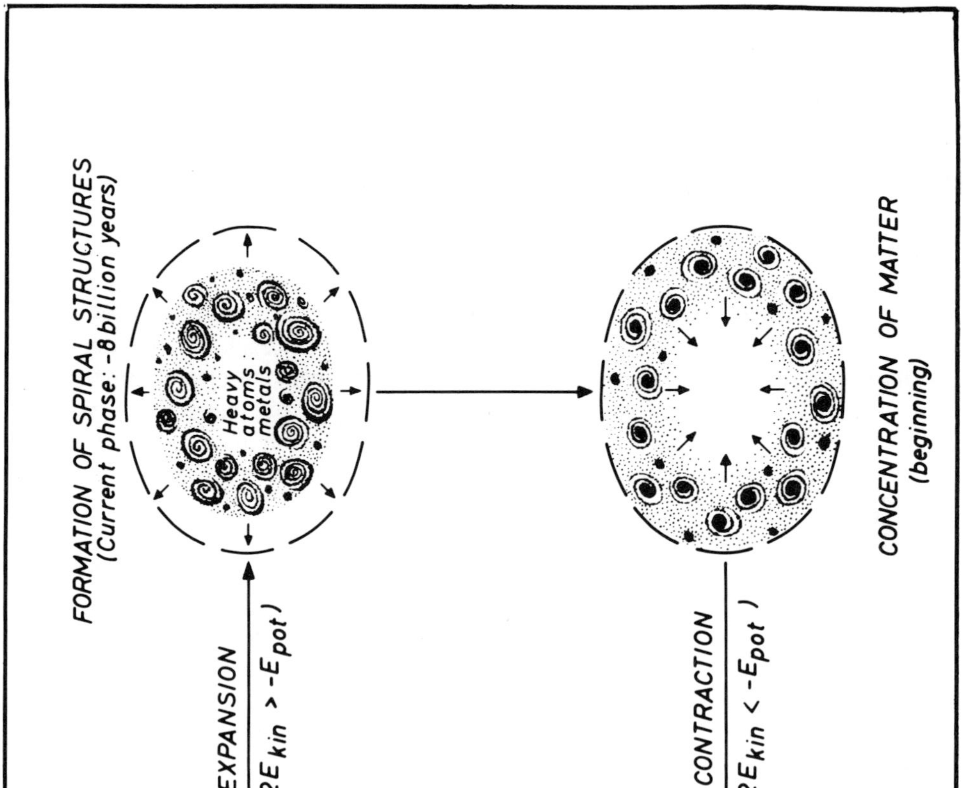
BIG BANG
Primal gas: e,p,n
Light atoms: H, He
EXPANSION
($2E_{kin} > -E_{pot}$)
FORMATION OF SPIRAL STRUCTURES
(Current phase: ~8 billion years)
Heavy atoms: metals
CONCENTRATION OF MATTER
(beginning)
CONTRACTION
($2E_{kin} < -E_{pot}$)
COLLAPSE
(beginning)

B. Pulsation model

According to this hypothesis, the cosmic forces of expansion, or the radiation pressure (in terms of Newtonian mechanics: kinetic energy, or E_{kin}), will come into equilibrium with those of mass attraction, or gravity (Newtonian mechanics: potential energy, or E_{pot}) in about 10^{10} years, i.e. the universe will then have reached the state of maximum expansion and minimum density (Dehnen 1976). In this condition first the stars and then the galaxies would begin to turn into black holes. The following overall cosmic contraction would then collect all the black holes, at first slowly and then ever more rapidly, into a single supermassive black hole. This would represent the entire universe. The result of such a cosmic collapse would be an enormous "heating" of matter, with all energy present as E_{kin}, which again would lead to a mighty explosion, the big bang. The resulting expansion would give matter outward linear momentum (**big bang phase;** Fig. 4). The mass would now expand in every direction as a volatile gas of elementary particles. During this phase, the force of gravity plays practically no role. However, as the expansion gradually slows down, gravity again becomes important as the driving force which leads locally to the formation of irregularities or turbulences. Such accumulations may initiate the long phase on the way to spiral formation. This is the phase in which the cosmic cycle of the last 10^{10} - $2x10^{10}$ years now appears to be. As cosmogenesis continues, the attractive forces gradually gain importance in respect to the expansive ones. The absolute value of twice the radiation pressure (E_{kin}) should again become smaller than the negative value of the gravitational force (E_{pot}), i.e. $2E_{kin} < -E_{pot}$. The point of maximum expansion of the radius of the universe and the minimum density of matter of the cosmos is reached again. Matter gradually begins to concentrate again, and the radius of the universe will decrease (**matter concentration phase**). This will finally lead to another aggregation of matter into a supermassive black hole (**collapse phase**) with a subsequent explosive release of rest energy, which will initiate the next cosmic cycle. The period of cosmic cycles can be approximately calculated from an **Einstein equation**. For the current cycle, a final age of $4x10^{10}$ - $8x10^{10}$ years is obtained.

V. Micro-mechanisms of cosmogenesis

Before we can discuss special mechanisms by which the external and internal changes in the form of the universe occur, we must consider the internal development of matter after the big bang. **Elementary particle** physicists believe that they can reconstruct these events exactly, starting from an unimaginably small fraction of a second after the big bang (e.g. Weinberg 1977, Barrow 1984). According to these authors, immediately after the big bang, the temperature of the universe was extreme, up to $2x10^{12}$ ^{o}K. At such temperatures matter is not atomic, but consists of a gas-like plasma of all possible free elementary particles. In this extremely dense original plasma, the elementary particles were in thermodynamic equilibrium, i.e. they collided sufficiently often with each other so that they were continuously materializing (as particles and antiparticles: 10^{-43} sec after the big bang) or annihilating (annihilation of proton-antiproton requires 10^{-6} sec; of electron-positron, 1 sec).The rapid expansion of the original plasma after the cosmic explosion must, however, have caused the temperature and density to decrease rapidly: within the first 30 sec to about $4x10^{8}$ ^{o}K, after one year to about $2x10^{6}$ ^{o}K and after a million years to $2x10^{3}$ ^{o}K. At this latter temperature, the number of collisions decreased to the point that some particles (e.g. neutrinos, photons) became "decoupled" from the others: they were no longer in equilibrium. They cooled independently to about 3 ^{o}K, the present temperature of the background radiation.

About 1-3 min after the big bang, however, density and temperature were high enough and consequently collisions of elementary particles frequent enough to enable the formation of hydrogen and helium **atoms** through fusion (1 electron + 1 proton -> 1 hydrogen atom; 2 electrons + 2 protons + 2 neutrons -> 1 helium atom); further fusion of the light atomic nuclei to form **heavy atoms** was presumably not concluded until 500,000 years after the big bang (e.g. 3 helium atoms -> 1 carbon atom; 4 helium atoms -> 1 oxygen atom; Fig. 5). At present, hydrogen makes up 80% of the matter in the universe , helium 19.9% , and the other elements make up only 0,1%.

The interactions between elementary particles due to gravitational instabilities, radiation pressure, electro-magnetism and electro-strong and electro-weak forces (compare p. 168, 170) are therefore responsible for the fact that after the big bang, the relatively high degree of disorder in the original plasma could be converted to the relatively well-ordered systems of atoms and thus also of celestial bodies.

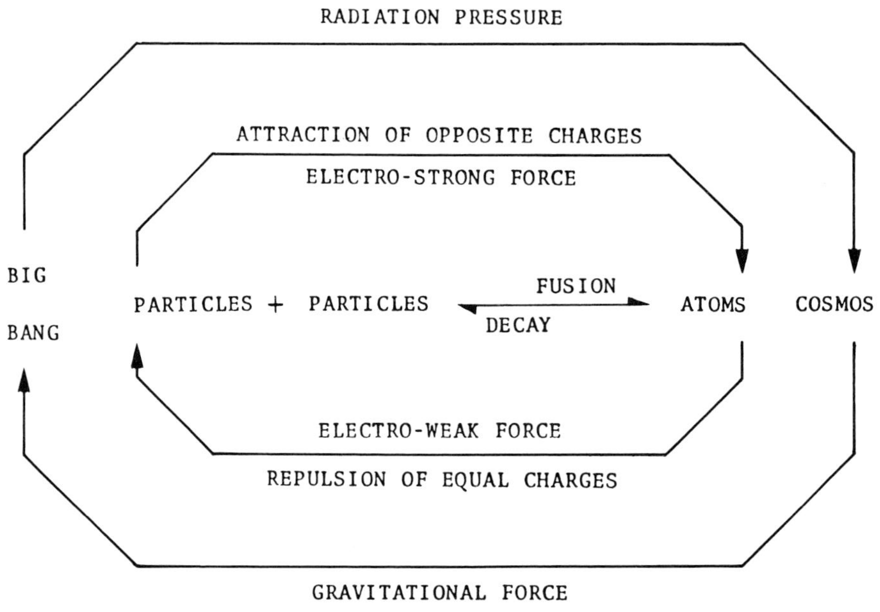

Fig. 5. Survey of important aspects of atomic decay and fusion processes (quantum mechanics) in the course of a cosmic cycle (Einstein's Theory of Relativity 1905, Hughes Pulsation Theory 1963, Hawking 1988).

VI. References

Barrow, J.D. and Silk, J. (1984): *The invisible universe*. New Scientist 103 (No. 1419) 28-30.

Davies, P. (1984): *Super force. Search for a grand unified theory of nature*. Heinemann, London.

Dehnen, H. (1976): *Die Struktur der Welt*. Umschau 76(7) 209-218.

Dicus, D.A., Letaw, J.R., Teplitz, D.C., Teplitz, V.L. (1983): *The Future of the Universe*. Scientific American 3 74-85.

Einstein, A. (1905): *Zur Elektrodynamik bewegter Körper*. Ann. Phys. 17 891-921.

Fritzsch, H. (1983): *Vom Urknall zum Zerfall. Die Welt zwischen Anfang und Ende*. Piper Verlag München, Zürich.

Hawking, St. (1988): *A Brief History of Time: From Big Bang to Black Holes*. Bantam Books, New York.

Hubble, E. (1936): *The realm of the nebulae*. Reprint 1982, Yale University Press New Haven.

Hughes, J.L. (1963): *Possible application of mass energy cycle on a micro- and macroscopic scale in the universe*. Nature 197 441-442.

Schwemmler, W. (1984): *Reconstruction of cell evolution: A periodic system*. CRC Press Boca Raton, Florida.

Silk, J., Szalay, A.S., Zel'dovich Y.B. (1983): *The Large Scale Structure of the Universe*. Scientific American 10 56-70.

Weinberg, S. (1977): *Die ersten drei Minuten. Der Ursprung des Universums*. R. Piper Verlag München.

Zimmermann, A. (1985): *Der Evolutionsbegriff in der Sicht der Kosmologie und der Physik*. In: Evolutionstheorie und Schöpfungsglaube. Engagement 4 330-336; Aschendorff Verlag Münster.

Chapter 3 Chemogenesis

I. Phenomenon

From our own "bird's eye view" through travel in space we know that the earth is a round celestial body which is divided into three regions: the lithosphere, hydrosphere and atmosphere. Comparisons with the corresponding regions of other planets, which are possible with modern techniques of cosmology, space exploration and geology, of paleochemistry and microbiology, give us information on the probable history of the origin of our planet. This insight in **geogenesis** allows us to infer the possible energetic and chemical conditions of the early stages of the earth. The question arises as to whether, under these conditions, the transition from molecules to the first living structure could have been possible. Such attempts are inseparable from the prior question, whether it is possible at all for the structural components of life, such as proteins, lipids, polysaccharides and nucleic acids, to form under the assumed conditions of the early stages of earth in purely abiogenetic fashion, i.e. without being influenced by living structures. At least some of the molecules must have been present, for the genesis of the first living structure from inorganic matter in the absence of these structural components is unthinkable.

II. Hypotheses

In the field of "chemogenetics", the study of prebiogenetic formation of the first structural components of life, rather contradictory hypotheses are presently being discussed. Most scientists postulate that **chemogenesis** is a natural, historical process of evolution, which occurred several billion years ago on earth. They believe that under the conditions of the early stages of earth, the inherent ability of matter to organize itself led almost necessarily, in a continuous process, to the prebiogenetic formation of proteins, lipids, saccharides and nucleic acids (among others; Kuhn 1972; Dose and Rauchfuß 1975; Eigen and Winkler 1976; Eigen 1983). However, a few other scientists express the contrasting opinion that the components of life could only have formed together with the first living systems by "innovation" in a kind of "creation" by external intervention (cf. Thürkauf 1984,

Vollmert 1985). The **hypotheses** of "**self-organisation**" and "**innovation**" are given in broad outline; explanations of the various other individual hypotheses on evolution of prebiogenetically relevant molecules will not be given since their basis is speculative rather than experimental. The data collected in the fields of cosmology, geology, paleochemistry and molecular biology will now be examined to see whether they support the hypothesis of self-organization or the contrasting one of "innovation". After that, the process of chemogenesis will be reconstructed on the basis of the more probable hypothesis. Obviously, this reconstruction will be one possible way of chemogenesis.

III. Data

A. Geogenesis

Direct observations by satellite and cosmological analysis of the origin of planets within and outside of our solar system have led to the general belief **that our earth formed about 5×10^9 years ago, as a final stage in the hierarchical process of cosmogenesis**, according to principles known to apply to small celestial bodies[1]. These principles say that in our solar system, mineral fractions condensed alongside with dust particles and gases, and thus the earth condensed into a glowing, liquid sphere as a result of the release of gravitational energy as heat. Due to conservation of angular momentum, slight rotation in the original gas cloud was transformed into a rapid spin of the planet during gravitational

---

[1] Explanations of the individual theories about geogenesis will not be given here.

Fig. 6. Schematic cross section of the **early** and **present**-day stages of the **earth**, suggesting the composition of the litho-, hydro- and atmospheres (not drawn to scale; Schwemmler 1984).

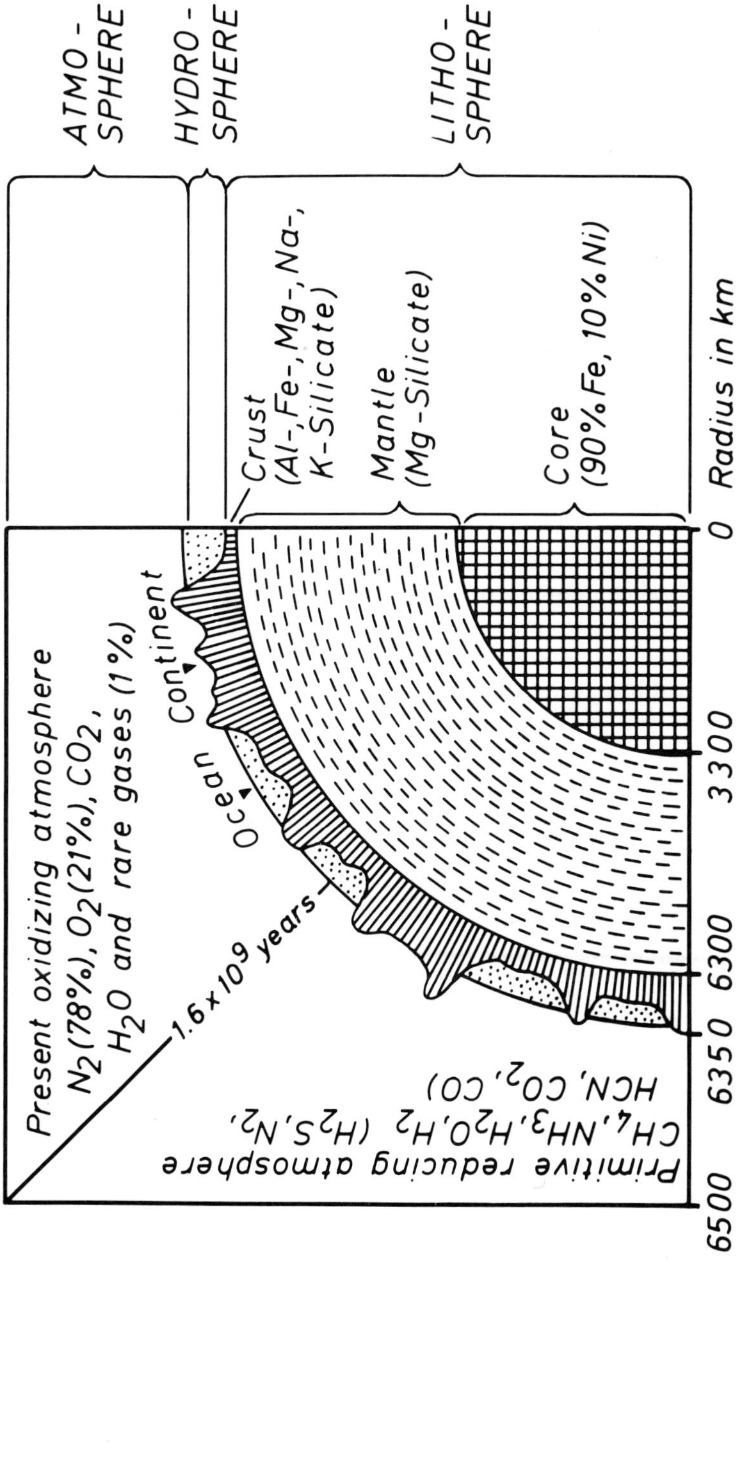
ATMO-SPHERE
HYDRO-SPHERE
LITHO-SPHERE
Present oxidizing atmosphere
N_2(78%), O_2(21%), CO_2, H_2O and rare gases (1%)
Primitive reducing atmosphere
CH_4, NH_3, H_2O, H_2 (H_2S, N_2, HCN, CO_2, CO)
1.6×10^9 years
Ocean
Continent
Crust (Al-, Fe-, Mg-, Na-, K-Silicate)
Mantle (Mg-Silicate)
Core (90% Fe, 10% Ni)
Radius in km
0
3300
6300
6350
6500

collapse[1]. The spin affected the formation of the **lithosphere**, **hydrosphere** and **atmosphere** as the earth cooled and these spheres in turn created the material boundary conditions for chemogenesis. Thus chemogenesis is a direct extension of cosmogenesis.

B. Boundary condition

At present there are no reliable data on the exact physical and chemical state of the primitive earth. However, geological studies of ancient earth layers give some basis for inferences about the chemical composition of the primitive atmosphere on earth. The oldest sediments, which are about $4x10^9$ years old, contain small **pyrite grains** (FeS_2). The position and characteristics of these grains indicate that they were exposed to the atmosphere for long periods. Since pyrites do not remain stable over a period of time in the presence of oxygen, i.e. they are easily oxidized, the primitive atmosphere cannot have contained significant amounts of **oxygen** (according to an estimate by Fahr 1984, there was about 1% O_2 due to photolysis of water). Studies of the atmospheres of other planets like Saturn and Jupiter, and of the gases released from volcanoes, have led to the conclusion that the **primitive atmosphere** must have been **reducing**, and composed primarily of the gases nitrogen (N_2), water vapour (H_2O), carbon dioxide (CO_2), ammonia (NH_3), hydrogen (H_2) and hydrogen sulfide (H_2S) (Schidlowski 1981; Fig. 6). The most recent terrestrial sediments which contain pyrite grains are about $2x10^9$ years old. On the other hand, the oldest terrestrial rocks which contain fully oxidized iron (Fe_2O_3) are about $1.5x10^9$ years old. This shows that the transition from the reducing primitive atmosphere to the present-day oxidizing atmosphere must have occurred between these two times. The chemical composition of the earth's mantle and the ionic composition (including pH) of the early seas, however, are thought by paleochemists not to have changed significantly, at least as far as the relative abundance of the elements is concerned.

The **energy conditions**, as well as the material ones, must have been significant for the course of chemogenesis. As there was **no protective ozone layer** at that time, short wave solar radiation would have been an important form of energy. However, earth energies like heat, lightning, radioactivity, cosmic radiation and occasionally supersonic waves caused by meteorite impacts, were surely also

---

[1] On the same principle, a spinning ice-skater spins more rapidly when she draws her extended arms in close to her body.

significant. In the final analysis, all the physical and chemical factors of the early stages of the earth, in their spatial and also temporal periodic distribution, may have had decisive effects on chemogenesis.

C. Simulation experiment

1. Prebio-monomer

In his experiment, which may be regarded today as a classic, **Miller** (1953) attempted to simulate some processes of **chemogenesis** in vitro under the deduced conditions of primitive earth (Fig. 7). The experiment simulated lightning by electric discharges, the gases CH_4, NH_3, H_2, and water vapour were added, resembling the primal atmosphere, and the early seas and rain were simulated by water in a circulating apparatus. The result of exposure of this gas mixture to the electric arc discharge was an enrichment in the aqueous phase of prebiogenetically synthesized organic molecules, the **prebio-monomers**. These included various amino acids and carboxylic acids, however only in trace amounts.

In simulation experiments, carboxylic acids can react with the trialcohol glycerol to form lipids. In experiments with gas mixtures containing only CO_2 and N_2, however, no amino acids were formed. Addition of energy in the form of ultraviolet light, ionizing radiation or heat at about 1000°C (e.g. volcanic heat) produced amino acids in mixtures containing CH_4 and NH_3 (for a review see Kaplan 1978). The yields of these simulation experiments were found to be about 1% of the carbon added as CH_4.

Other simulation experiments demonstrated the synthesis of the starting materials for **nucleic acids**: sugars, nucleo-bases and phosphates. Ultraviolet irradiation of pure HCN solutions led to the formation of the purine bases, adenine and guanine, and heating of aqueous urea and cyanoacetylene or acrylonitrile solutions led to the synthesis of cytosine and uracil (for a review see Dose and Rauchfuß 1975). Yet these results do not give much evidence for biogenesis because pure solutions of HCN, urea, cyanoacetylene and acrylonitrile are unlikely to have been present on the prebiotic earth.

The prebiotic formation of the **sugars**, triose, tetrose, pentose and hexose, has been achieved experimentally, whereas there is no agreement on whether riboses and deoxyriboses are also formed prebiogenetically (for a review see Dose and

Rauchfuß 1975). Formaldehyde has been detected in interstellar space. The prebiotic synthesis of nucleotides, the components of nucleic acids, was observed after irradiation of an aqueous solution of bases, ribose (a pentose) and phosphates (Ponnamperuma et al. 1963); the phosphate acted as catalyst. When in similar experiments free phosphoric acid was used at 20°C, mono-, di- and triphosphates of the purine and pyrimidine bases were obtained in yields greater than 1% of the starting materials (Waehneldt and Fox 1967). However, the evaluation of these experiments for biogenesis is controversial since the prebiotic existence of ribose has not been proven up to now.

2. Prebio-polymer

It could also be shown that amino acid mixtures, when heated to about 200°C, could be made to condense (Fox and Waehneldt 1968). The amino acid composition of these prebiogenetic "**protenoids**" or **prebio-polymers** has been found not to differ significantly from that of "modern" biogenetic proteins, as had been determined by hydrolysis of the protenoids. The molecular weights of such protenoids are between 4,000 and 10,000, which corresponds to the weight of small protein molecules. It could also be shown that under simulated prebiotic conditions, the sequence of amino acids in the peptide chains was not strictly a matter of chance but that amino acid combinations with catalytic abilities (α-peptide bonds) occurred at a higher relative frequency. In addition, the

---

Fig. 7. Experimental set-up used by **Miller** (1953) to study the formation of organic compounds in a simulated primitive reducing atmosphere subjected to arc discharges (adapted from Fox and Dose 1972). The gases NH_3, CH_4, H_2, and water vapour circulated through a system of tubes and energy was applied in the form of electric discharges. The experiment ended after several days, and the content of the trap was analyzed.

As an alternative to Miller's investigations, recent experiments carried out by **Freund** (1981: Atomarer Kohlenstoff; Nachrichten aus Chemie, Technik und Laboratorium; Ges. Dsch. Chem. 29 301) raise the possibility of the prebiotic origin of carbohydrates by way of purely inorganic reactions, i.e. from reactive carbon atoms in crystals.

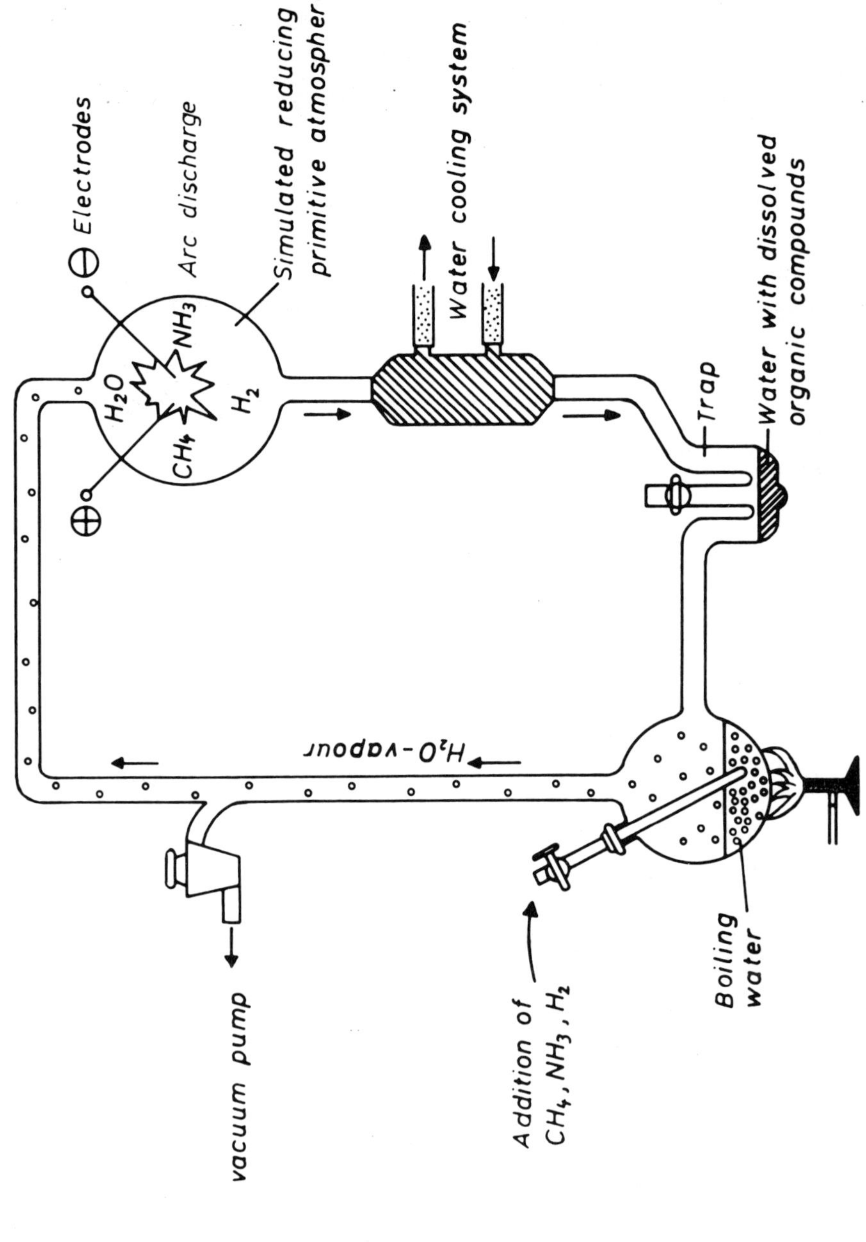

Electrodes
Arc discharge
Simulated reducing primitive atmosphere
H_2O
NH_3
CH_4
H_2
Water cooling system
Trap
Water with dissolved organic compounds
H_2O-vapour
vacuum pump
Addition of CH_4, NH_3, H_2
Boiling water

quantitative yields of certain polymers were much higher than a statistical random distribution would predict(Reinbothe and Krauss 1982).

The formation of **prebiogenetic polynucleotides** has been reported when aqueous solutions containing polyphosphates and nucleotides were treated with X-rays or ultraviolet light (for a review see Dose and Rauchfuß 1975). About half these products were linked by phosphodiester bonds, although the 3'-5' bonds, which are exclusively present in modern biogenetic systems, have been rarely observed. Due to a questionable mode of analysis, these results are doubted by many experts. Inue and Orgel (1983) observed that in the presence of natural metal catalysts (e.g. zinc ions), RNA molecules of more than 100 nucleotides can be synthesized in which certain RNA nucleotides are preferentially incorporated.

IV. Reconstruction

A. The problem

Although we do not exactly know what the physical-chemical boundary conditions in the early state of earth were like, we can assume that the raw materials for chemogenesis were most probably the gases of the primal atmosphere, and that metal ions acted as catalysts in the aqueous phase. We may also assume that adequate prebiotic sources of energy were available.

Various laboratory experiments, designed to simulate the presumed conditions on the primitive earth, appear to demonstrate that biologically relevant basic compounds of life, i.e. amino acids and carboxylic acids - with the exception of the nucleotides - can, in principle, be formed purely prebiogenetically. However, it is a matter of heated debate whether polycondensation of these prebio-monomers, again abiogenetically, could have produced the prebio-polymers, proteins, lipids, polysaccharides and nucleic acids, under prebiotic conditions.

Just as the hierarchical series of organisation of cosmic systems suddenly stops with the formation of satellites of planets, critics argue that after the formation of prebio-monomers the causal chain of evolution of the cosmic primal gas reaches an abrupt caesura. It is objected that all polymerizations so far achieved in simulation experiments follow the known laws of polymer chemistry, thermodynamics, stoichiometry and statistics, but that these are irrelevant for prebiotic conditions, and thus do not admit the formation of prebio-polymers in the "primal soup". The essentials of the objections are as follows (Vollmert 1985):

1. If the water produced by polycondensation is not largely removed from the reaction mixture, the reverse reaction will occur, i.e. the polymer will be hydrolysed again. It is unclear how this problem can be solved in an aqueous "primal soup".

2. When a polymer chain grows, various types of monomers may compete for positions in the chain. If chemically varied bifunctional monomers are present, this necessarily leads to mixed polymers; when monofunctional monomers are incorporated, the chain breaks off. According to the simulation experiments, mixtures of various mono- and bifunctional monomers must have been present in the "primal soup" in widely varying concentrations. How could pure prebio-polymers have been formed from such uneven mixtures?

3. If polycondensations are to be achieved in vitro, the reacting monomers must be present in certain ratios, degrees of purity and in minimum amounts. Here again, it is an open question whether and how such conditions could have been fulfilled in the "primal soup".

On the whole it seems that the direct application of the laws of thermodynamics, stoichiometry and statistics of the closed systems of polymer chemistry to the basically open system of the prebiotic "**primal soup**" is questionable, if not actually inadmissible. The reactions in the "primal soup" are exposed to very different conditions, e.g. the effects of a periodically changing environment (day/night, high/low tide, summer/winter), of special catalysts and of molecular matrices (for extensive discussion, see Eigen 1983). The problem of removing the water of condensation from the reaction mixture has also been solved, in principle (see next chapter, p. 31).

In an evaluation of the **Miller** experiment and the question of the origin of nucleotides and chirality, i.e. the selective enrichment of optically pure D- or L-forms, **Dose** (1987) finds the following basic difficulties in the concepts of primal atmosphere and "primal soup":

1. It is now generally believed that the primitive atmosphere of the earth was formed from volcanic outgassing and consisted mostly of CO_2, H_2O, and N_2, rather than of H_2, CO and H_2S. Thus it probably never contained the large amounts of free H_2, NH_3 and CH_4 which were used by Miller-Urey in his first experiments.

2. It also seems certain that the reactive intermediates which presumably formed in the primitive seas must rapidly have reacted further to ill-definable polymers. In light of the low yields of prebiogenetic organic molecules obtained in the simulation experiments, it is therefore not easy to imagine today how life could have arisen in such a thin "primal soup", and this would cast doubt on the entire concept of the "primal soup".

3. The main products of the Miller *experiments seem to be more closely related to coal tar or to the hydrocarbons in carbonaceous meteorites than to the molecular inventory of prebiotic minimum organisms. Furthermore, in the second generation of simulation experiments which followed the Miller experiments, relatively high concentrations of starting materials were used; no mechanism has been suggested which would account for a prebiotic accumulation or temporary stabilization of the precursor molecules, most of which are reactive. In 1982* Cairns-Smith *summarized the critical objections to the Miller experiment and biogenesis. The alternative hypothesis on the mineral origin of life which is based on Cairns-Smith's considerations, is presently still being verified by a detailed series of tests.*

4. Finally, there are still no unequivocal results of laboratory simulations of the prebiotic formation of polynucleotides capable of replication, which would require the accumulation of optically pure D- or L-forms.

Even though chemogenesis is, up to now, the weakest link in the chain of arguments of the evolution theory - as there is no definite experimental evidence for the formation of replicative nucleic acids of pure chiral systems[1] under prebiotic conditions - **the hypothesis of self-organisation on the whole seems more plausible than the innovation hypothesis.** According to **Dose** (1987) the former is supported mainly by "the abundance of bio-elements in the universe, the formation of a significant variety of organic molecules by simulated prebiotic reactions and in interstellar systems, and the thermodynamic plausibility of evolution and self-organisation". Therefore, the model of chemogenesis developed here is a self-organizing process; it is presented in the form of a short speculative sketch which, in this form, is not supported by all "chemogeneticists".

B. Self-organization model

The self-construction of prebio-monomers in the early state of earth could have begun about 4.5×10^9 years ago (Fig. 8). It might have been initiated by formation

---

[1] Mirror image symmetry is the steric structure of molecules with asymmetrical atoms (e.g. a carbon atom with four different ligands); adequate and necessary condition for the occurrence of optical activity, i.e. the event of polarized light diverting from its plane.

of **radicals** or **ions** in the primitive atmosphere through the effects of different form of energy:

$$A : B \dashrightarrow A^{\cdot} + {}^{\cdot}B \text{ (radicals)}$$

$$A : B \dashrightarrow A^{-} + B^{+} \text{ (ions)}$$

Such radicals and ions could have reacted immediately to form intermediates like hydrogen cyanide (HCN), formaldehyde (HCOH), acetylene (HC≡CH), ethylene ($H_2C=CH_2$), formic acid (HCOOH) etc. Comparable intermediates have been observed in interstellar space. The intermediates might have been linked in many ways to form the prebio-monomers: sugars, formic acid, carboxylic acids and possibly nucleosides. These small molecules may have become enriched by selective periodic processes such as adsorption on and desorption from active surfaces (foam, mud, clay and minerals of e.g. zeolithes and montmorillonites), by processes of filtration, sedimentation, recrystallization or by drying out and redissolving in the "primal soup". In this way, a variety of reaction spaces would have arisen which differed spatially and temporally, but which contained fairly pure monomers of a given type.

As the next step, the partially selected prebio-monomers in such areas could have reacted further to form pure di- to oligomeric molecules. The water released by these condensation reactions might either have evaporated (dry condensation on hot lava) or may have been chemically bound, perhaps through derivates of hydrogen cyanids (or minerals):

R^1-OH	+ H-R^2	+ H_2N-CN	---> R^1-R^2	+ H_2N-CO-NH_2
Monomer	+ Monomer	+ Cyanamide	---> Dimer	+ Urea

In further condensation to pure prebio-polymers of protein, polysaccharides, lipids and maybe also nucleic acids, certain metal catalysts, periodic oscillations in the environment (day/night, high/low tide, summer/winter) and the matrix effects of prebio-oligomers may also have been decisive influences. Such factors are thought to have contributed to the formation of prebiomers in which the sequences were not strictly random (Kuhn 1972). In the case of ribonucleic acid (RNA), a matrix effect possibly appears; after a certain size has been reached, the molecules tend to replicate and lengthen. In other words, they do not arise randomly, but a sequence which is already present is copied, and in some cases multiple copies are lined up in a row, so that longer sequences can arise. In addition, matrix-directed synthesis of RNA does not represent an equilibrium system, because intervening selection enters and has enormous effects on the probability of distribution (Eigen 1983).

Here too, an explanation for those mechanisms may be found which eventually led to prebiomers containing only one kind of optical isomer. **Follmann** (1981), for example, suggests that the left-handedness of natural ß-radiation might lead to enrichment of one optical isomer. For example, D-tyrosine was more rapidly destroyed by polarized ß-rays from ^{90}Sr (Strontium) decay than was L-tyrosine (Dose and Rauchfuß 1975). Which of the two stereoisomers was finally selected might depend either on chance (Eigen and Winkler 1976) or on asymmetric radiation effects (Follmann 1981), depending on the mechanisms of selection or maybe as a consequence of circular polarized radiation from the "black holes" (Dose 1987, personal communication).

V. Micro-mechanisms of chemogenesis

It is evident that chemogenesis is closely connected to cosmogenesis. If the "condensation" into stars and planets converted gravitational potential energy to heat, then just such heat from the sun and earth would now promote the conversion of small molecules into macromolecules in the course of chemogenesis. The tendency to form more and more complex molecules - called **self-organisation of matter** - is a property of matter itself which it expresses under suitable conditions. The ability of matter to become organized depends on the chemical properties of the elements and molecules: formation of atoms, bond formation (ionic, polar covalent, and nonpolar covalent bonds), intermolecular

---

Fig. 8. Schematic representation of the primitive **earth as a "chemical laboratory"** for the synthesis of prebio-monomers, prebio-oligomers, and prebio-polymers* (according to Kaplan 1978).

* Prebiogenesis highly questionable: poor possibility of synthesis, high rate of decomposition.

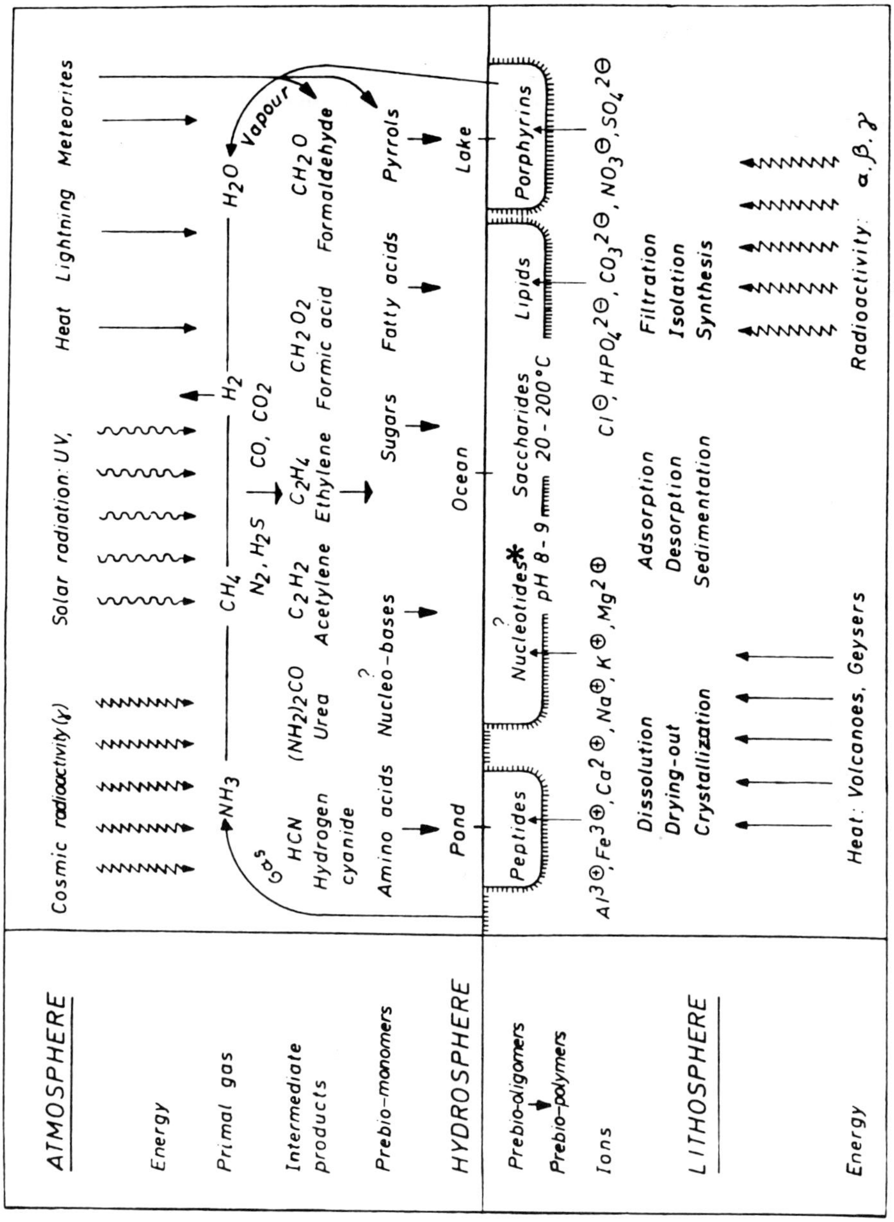
ATMOSPHERE
Energy
Primal gas
Intermediate products
Prebio-monomers
HYDROSPHERE
Prebio-oligomers
Prebio-polymers
Ions
LITHOSPHERE
Energy
Cosmic radioactivity (γ)
Solar radiation: UV,
Heat
Lightning
Meteorites
NH3
CH4
H2
H2O
N2, H2S
CO, CO2
Gas
Vapour
HCN Hydrogen cyanide
(NH2)2CO Urea
C2H2 Acetylene
C2H4 Ethylene
CH2O2 Formic acid
CH2O Formaldehyde
Amino acids
? Nucleo-bases
Sugars
Fatty acids
Pyrrols
Pond
Ocean
Lake
Peptides
? Nucleotides*
Saccharides
Lipids
Porphyrins
pH 8 - 9
20 - 200°C
Al3⊕, Fe3⊕, Ca2⊕, Na⊕, K⊕, Mg2⊕
Cl⊖, HPO42⊖, CO32⊖, NO3⊖, SO42⊖
Dissolution
Drying-out
Crystallization
Adsorption
Desorption
Sedimentation
Filtration
Isolation
Synthesis
Heat: Volcanoes, Geysers
Radioactivity: α, β, γ

forces (e.g. hydrophoby, hydrophily, dipole-dipole forces and Van der Waal's forces, complex formation), spatial orientation (secondary, tertiary, quarternary structure), aggregate states, energy uptake and release, and so on. When they are synthesized, molecules store a large amount of potential chemical energy and for this reason are unstable; they would not become enriched in an unchanging environment. Herein lies the importance of a **periodically changing environment** in which the temperature, and thus the reactivity of molecules, changes. There is a periodic drying out and redissolving in the course of tidal and seasonal rhythms as well as periodic exposure of molecules to atmospheric energy sources, from which they are removed again by rain and tides. Thus the changing environment effects a whole complex of interacting processes, which are constantly repeated, and which constantly drive the chemical systems from one extreme of conditions to the other (**energy dissipation**). In this way they achieve the necessary complexity (in contrast to the homogeneous conditions of polymerization in a closed system in vitro) so that selection can have an effect. It is proposed that the unbroken chain of ever more complex chemical systems, up to the precellular mixed molecular aggregates, was generated by the interaction between the mechanisms of self-organization and selection in the course of chemogenesis (Kuhn 1972).

VI. References

Cairns-Smith, A.G. (1982): *Genetic takeover and the mineral origins of life.* Cambridge University Press, New York.

Dose, K. (1987): *Präbiotische Evolution und der Ursprung des Lebens - ein kritischer Rückblick.* Chemie in unserer Zeit 6 177-185. .

Dose, K., Rauchfuß, H. (1975): *Chemische Evolution und der Ursprung lebender Systeme.* Wiss. Verlagsgesellschaft, Stuttgart.

Eigen, M., Winkler, R. (1976): *Das Spiel. Naturgesetze steuern den Zufall.* Piper Verlag München.

Eigen, M. (1983): *Entstehung des Lebens. Ein Ereignis zwischen naturgesetzlichem Zwang und historischer Einzigartigkeit.* Natur 3 68-77.

Fahr, H.J. (1984): *Am Anfang war die Sonne blau.* Bild der Wissenschaft 2 76-82.

Follmann, H. (1981): *Chemie und Biochemie der Evolution.* Quelle und Meyer, Heidelberg.

Fox, S.W. and Dose, K. (1972): *Molecular evolution and the origin of life.* Freeman, San Francisco.

Fox, S.W. and Waehneldt, T.V. (1968): *The binding of basic protenoids with organismic or thermally synthesized polynucleotides*. Biochim. Biophys. Acta 160 239-245.

Inue, T. and Orgel, L. (1983): *A nonenzymatic RNA polymerase model*. Science 219 859-862.

Kaplan, W. (1978): *Der Ursprung des Lebens. Biogenetik, ein Forschungsgebiet heutiger Naturwissenschaft*. 2nd edition; Georg Thieme Verlag, Stuttgart.

Kuhn, H. (1972): *Selbstorganisation molekularer Systeme und die Evolution des genetischen Apparates*. Angew. Chemie 84(18) 838-860.

Miller, S.L. (1953): *A production of amino acids under possible primitive earth conditions*. Science 117 528-529.

Orgel, G.E., Crick, F.H.C. (1980): *Selfish DNA: the ultimate parasite*. Nature 284 604-607.

Ponnamperuma, C., Sagan, C. and Mariner, R. (1963): *Synthesis of adenosintriphosphate under possible primitive earth conditions*. Nature 199 222-226.

Reinbothe, H. and Krauss, G.-J. (1982): *Entstehung und molekulare Evolution des Lebens*. Fischer Verlag Jena.

Schidlowski, M. (1981): *Die Geschichte der Weltatmosphäre*. Spektrum der Wissenschaft 4 17-27.

Schwemmler, W. (1984): *Reconstruction of cell evolution: A periodic system*. CRC Press, Boca Raton (Florida).

Thürkauf, M. (1984): *Wissenschaft schützt vor Torheit nicht*. Jordan Verlag Zürich.

Vollmert, B. (1985): *Das Molekül und das Leben. Vom makromolekularen Ursprung des Lebens und der Arten*. Rowohlt Verlag Hamburg.

Waehneldt, T.V. and Fox, S.W. (1967): *Phosphorylation of nucleosides with polyphosphoric acid*. Biochim. Biophys. Acta 134 1-8.

Chapter 4 Biogenesis

First Subchapter Precyte

I. Phenomenon

A final criterion for life, which is met only by cells, is the capacity for autonomous self-construction. **Monod** (1970) coined the phrase "**autonomous morphogenesis**" to describe this phenomenon. It means the forming of new living beings from parents and in cooperation with the environment; it is always connected with metabolism, therefore with the taking up, reorganizing and excreting of matter. These processes in turn require cytoplasm and a physiological barrier, i.e. at least a semipermeable membrane. Self-reproduction also means passing on characteristics from one generation to the next, the characteristics resulting from coincidental, rare and unintentional hereditary changes (mutations). The minimum requirements for autonomous morphogenesis are thus a semipermeable membrane, a metabolically active protoplasm, and a genetic apparatus which is capable of self-reproduction. As these conditions are met by cells, the phenomenon "life" is tightly bound to the existence of a cell of some sort. A cell therefore is the smallest unit of all living creatures, capable of living and reproducing autonomously.

These considerations point to the minimum structures which must have formed and to the minimum processes which must have occurred in the transition from molecules to the first cell in the "phylogenetic tree", the so-called "**ur-cell**"(from German *Urzelle*, meaning the "original" cell). Some of the conjectures about the origin of ur-cells - and of life itself - are presented below (Fig. 9).

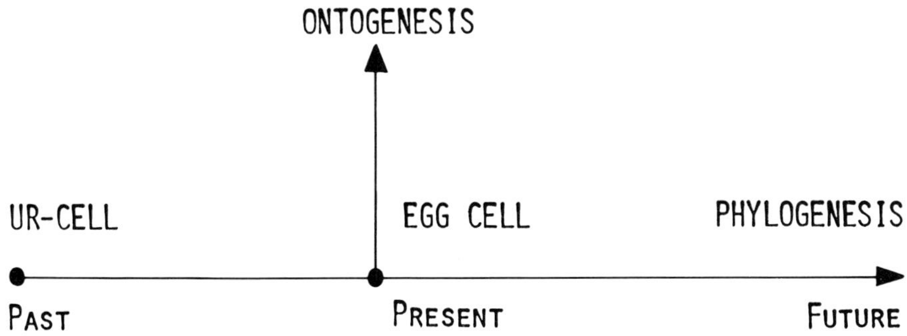

Fig. 9. Biogenetic derivation* of the ur-cell. As a rule there is an individual cell at the beginning of individual development , likewise, according to the "fundamental biogenetic rule", there should have been an individual cell, called the ur-cell, at the beginning of the phylogenetic development of life on earth.

* **Ernst Haeckel** (1834-1919) established the rule that ontogeny (individual development), to a certain degree, is a brief recapitulation (repetition) of certain embryonic stages in phylogeny (development or evolution of phyla; for details see p. 140).

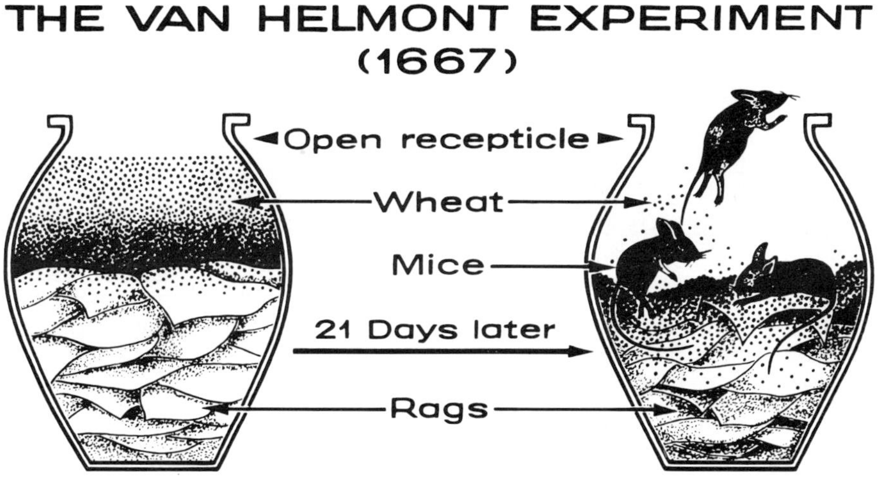

Fig. 10. According to **van Helmont** (1667), sweaty underclothes and wheat which had been filled into an open container were transformed into mice following the formation of a "ferment". Particularly remarkable is the observation that mice of both sexes appear, which are even able to reproduce fertile descendents (according to Dose and Rauchfuß 1975).

II. Hypotheses

Human beings have always been interested in the origin of life. In antiquity and in the Middle Ages, the teaching of **Aristotle** (382-322 BC) that life could arise from the non-living, was universally accepted. Not only plants, but animals as well, were thought to arise from dew, mud and/or manure by **spontaneous generation** (generatio spontanea). As late as the 17th century, the highly respected physician **van Helmont** published a scheme for generating mice from wheat and rags, as a sort of "recipe" (Fig. 10), without anyone seriously disputing it.

The careful experiments of **Louis Pasteur** (1822-1895) brought the first scientific proof that life is never formed anew spontaneously, at least not on the earth as we now know it. In a famous speech given at the Sorbonne in Paris (1864), Pasteur commented: "The doctrine of spontaneous generation will never recover from the fatal blow it has received from this experiment". **William Preyer** then summarized the new understanding in 1873 in the phrase "**omne vivum e vivo**" ("all life comes from life"). It remained for the scientific posterity of this generation to deal with the paradox that at the beginning of biological evolution, life must have developed from the non-living. One logically possible way out of the dilemma was first formulated in 1908 by **Svante Arrhenius** and later developed further by **Hoyle** and **Wickramasinghe** (1982), and then by **Crick** and **Orgel** (1983): the **panspermia hypothesis**.This hypothesis is that life arrived on earth in the form of spores from space. However, this does not solve the problem of the origin of life, but only displaces it into space.

Finally, **simulation experiments**, including those of **Stanley Miller** (1953), gave a scientific basis for the hypothesis that under the supposed conditions of the primitive earth, non-living matter became organized into living matter (Fig. 11). According to this young branch of science, **biogenetics**, molecules like amino acids, sugars, fatty acids and maybe nucleo-bases were formed under certain physical-chemical conditions, from the gases of the primitive atmosphere. Out of these macromolecules of proteins, polysaccharides, lipids and maybe nucleic acids formed (not yet verified experimentally).

Fig. 11. Schematic drawing of the primitive **earth as a "chemical laboratory"** for the formation of **minimum organisms** from prebiotically formed mixed aggregates (adapted from Kaplan 1978). Chemogenesis is the phase coupling cosmic and biological evolution.

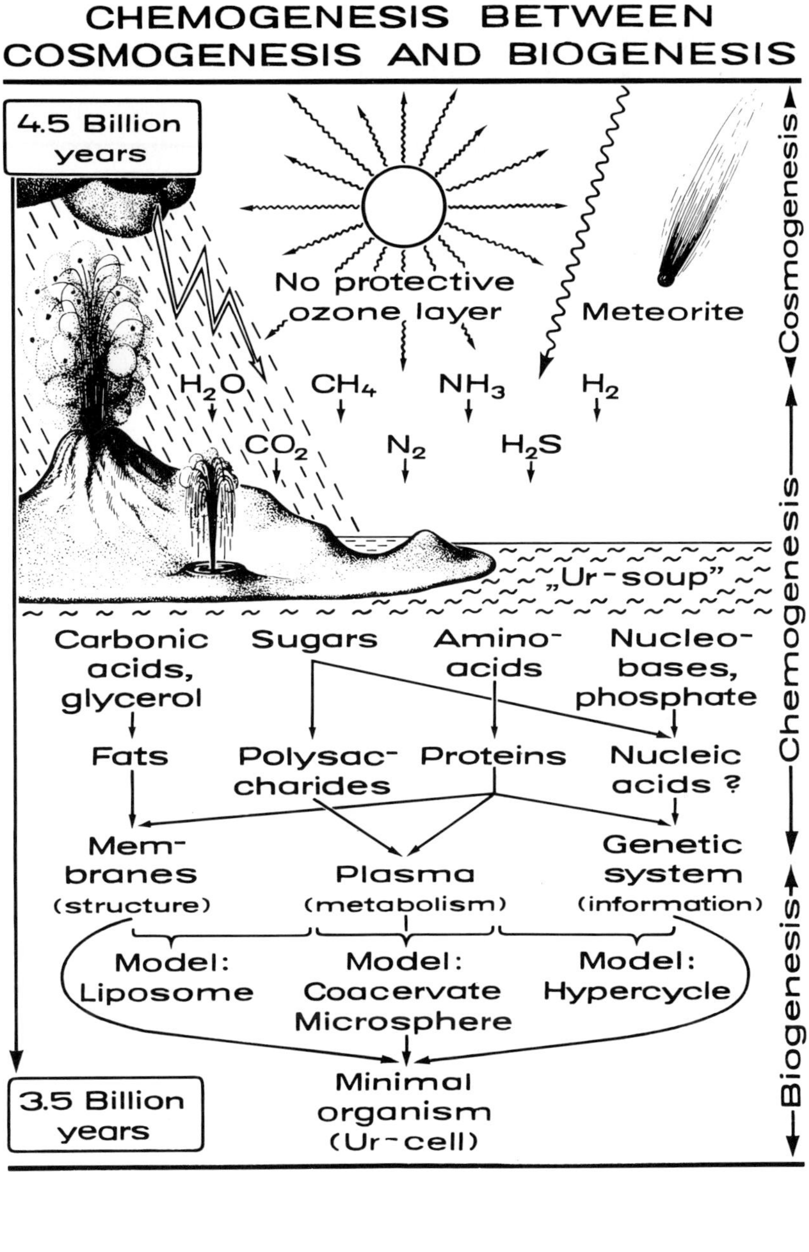
CHEMOGENESIS BETWEEN COSMOGENESIS AND BIOGENESIS
4.5 Billion years
No protective ozone layer
Meteorite
H_2O
CH_4
NH_3
H_2
CO_2
N_2
H_2S
„Ur-soup"
Carbonic acids, glycerol
Sugars
Amino-acids
Nucleo-bases, phosphate
Fats
Polysac-charides
Proteins
Nucleic acids ?
Mem-branes (structure)
Plasma (metabolism)
Genetic system (information)
Model: Liposome
Model: Coacervate Microsphere
Model: Hypercycle
Minimal organism (Ur-cell)
3.5 Billion years
Cosmogenesis
Chemogenesis
Biogenesis

These in turn were the basis for the mixed aggregates of membrane, protoplasm and genetic apparatus, which arose and became integrated into the minimum structure of the **ur-cell** due to the inherent self-organizing tendencies of these molecules (for a review, see Dose and Rauchfuß 1975).

At present there are two main hypotheses concerning the order of origin and integration of membrane, protoplasm and genetic apparatus; these can be reproduced here only in schematic fashion (for a review, see Schwemmler 1984). The **multi-step hypothesis** postulates a stepwise development of the ur-cell, starting from **a single** mixed aggregate of prebiogenetic proteins and nucleic acids, which could have been coupled to one another in a hypercycle (Eigen 1971). In the course of the hypercyclic coupling, synthesis of nucleic acids and proteins may proceed by means of a series of specific replicases with a double function. Only later would protoplasm and plasma membrane have evolved under genetic control. The driving mechanism of evolution, besides mutation, might have been an external periodicity or the day/night or summer/winter rhythm (Kuhn 1976). A competing view is the **multi-hit hypothesis**, which is the supposition that the ur-cell was started by simultaneous integration of **several** mixed aggregates as prebiogenetic precursors of the membrane, protoplasm and genetic apparatus. Accordingly **Kaplan** (1982) expresses the simpler, more probable opinion, that double systems with only one replicase for all genes were formed first. The essential difference between the two working hypotheses lies in the order of appearance of the **cell membrane** and the **genetic apparatus**. While the multi-hit hypothesis assumes that an prebiogenetic semipermeable membrane or limit of some sort[1] must have been present before the formation of the genetic apparatus, the multi-step hypothesis assumes that a membrane could only have formed biogenetically after the genetic apparatus had arisen. The novel hypothesis of **Cairns-Smith** (1982) about the mineral origin of life is an interesting train of thought, but still lacks experimental support.

---

[1] A membrane is one of the ways an individual can retain nucleic acids and proteins which would otherwise be lost.

Another possibility is a small lump of gel, made up of nucleic acids and proteins, which is also semipermeable, and therefore retains nucleic acids and proteins, and allows nutrients (amino acids, nucleotides) to find their way in.

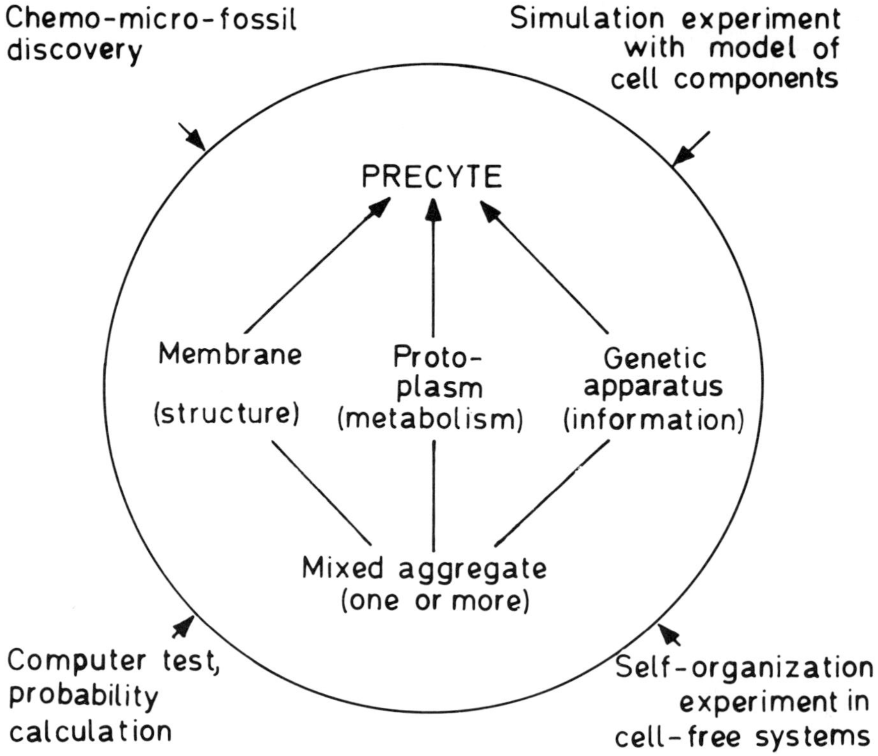

Fig. 12. Fossils, analytical data and hypotheses concerning the evolution of ur-cells.

III. Data

Of course, one may dismiss such hypotheses as pure speculation, but there are a number of outstanding findings which greatly increase the probability of the one or the other postulated evolutionary step (also see abstracts on "Origin of Life", 1986; Fig. 12).

A. Fossil discovery

Direct evidence of prebiomers, in the form of **chemical fossils**, is difficult to obtain, due to their instability and the great degree of geochemical change in the earth since its formation. The few unambiguous discoveries of substances produced by decomposition do suggest, however, that the prebiogenetic macromolecules (proteins, lipids, carbohydrates and probably nucleic acids) necessary for the evolution of the ur-cell must have been present simultaneously at the beginning (for a review, see Dose and Rauchfuß 1975; Kaplan 1978). These could have reacted further to form mixed aggregates even about $4x10^9$ years ago, since the oldest, though not undisputed **cellular fossils** date from about $3.7x10^9$ years ago. These fossils were discovered in the **Isua-mica-metaquartzite** layer of Greenland (Nagy 1976). If they turn out to be chemical artefacts, as supposed by paleochemists, the next oldest possible microfossils will be the 10-30 μm spherical cells from the $3.1x10^9$ years old Fig Tree formation of South Africa (cf. Fig. 13). A more definite characterization of this finding as membrane-enclosed ur-cells is not possible in this context (for a review, see Dose and Rauchfuß 1975). Paleontological studies of fossils still do not answer the question of prebiogenetic formation of membranes, at least for the present.

B. Cell component simulation

The oldest in vitro model of possible ur-cells is the **coacervate**, introduced by **Oparin** (1924, 1968). The original coacervates were mixed aggregates of biogenetic macromolecules, e.g. carbohydrates and proteins, which were obtained by dissolving these substances in salt solutions at certain concentrations. Controlled changes in pH then cause small but unstable droplets to precipitate (Fig. 14a). The salt concentrations in the droplets may be up to one hundred times as great as in the surrounding solution. Simple metabolic processes, e.g.

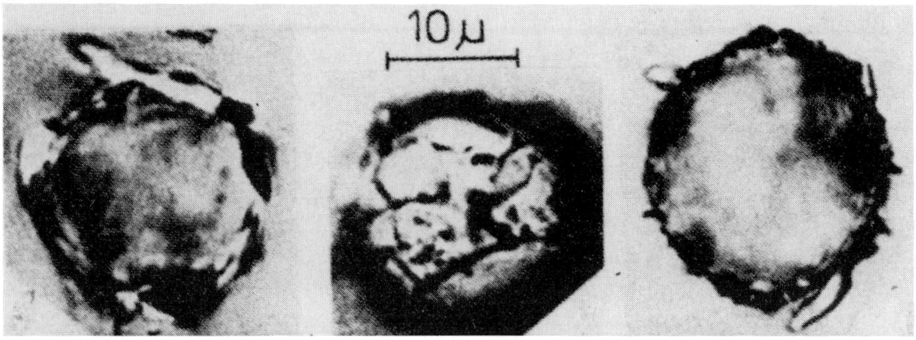

Fig. 13. Supposedly oldest substantiated **cellular fossil** discoveries from the $3.1x10^9$ years old Fig Tree formation in South Africa (according to Pflug et al. 1969).

growth by synthesis of starch, can be simulated in these coacervates (Fig. 14 b). They are also comparable to protoplasm with respect to their colloidal chemistry. However, there is opposition to Oparin's opinion, that coacervates are models for ur-cells. At best, they may represent models for pre-organelles, i.e. for precursors of **cell component systems** such as the protoplasm, since they lack a stable limiting membrane (Fig. 14c). However, it remains Oparin's contribution to regard biogenesis as a necessarily physical-chemical process.

Sydney Fox (1965,1972) continued and expanded upon Oparin's experimental work. Among other things, Fox heated dry mixtures of amino acids to 200°C and dissolved the resulting artificial proteins (also called **protenoids**) in seawater. This resulted in the formation of approximately cell-sized spheres, which he named **microspheres**, some of which were enclosed in semipermeable membranes and were capable of division, growth and motion including hydrolysis, decarboxylation, amination, deamination and oxidation/reduction. They can be used to study regular interactions between proteins and nucleic acids which are reminiscent of ribosomal reactions (cf. Barbieri 1981). In microspheres, polycondensation of proteins on nucleotides was carried out with and without inorganic catalysts. These so-called **simultaneous syntheses** indicate reactions which are stoichiometric in nature (Krampitz et al. 1969, Paecht-Horowitz et al. 1970; Fig. 16). These experiments also strengthen the idea that the present genetic code probably has a prebiotic basis in stereochemical, kinetic and thermodynamic parameters (for a review, see Kaplan 1978, 1982). Nevertheless, stable microspheres are not, contrary to the claims of **Fox**, a convincing model for the ur-cell because they lack a system which could coordinate the self-replication and mutation of protoplasm, membrane and genetic apparatus in an ur-cell. Rather, they may serve as a model for precursors of the **cell component systems** of protoplasm or genetic apparatus. On the other hand, Fox's conclusion that the further development to a genetic apparatus could only have occurred in a limited region, is convincing.

C. Evolution experiment in vitro

A further possibility for studying the evolution of ur-cells is offered by cell-free systems in vitro. The regularities of association of macromolecules have also been studied in this way; however, so far only **biogenetic** macromolecules have been studied. Stable protein-lipid complexes with regular structures and specific transport characteristics, for example, are obtained in vitro simply by mixing the molecular components. Depending on the experimental procedure, either monolayers or bilayers form which may serve as **membrane models**. By

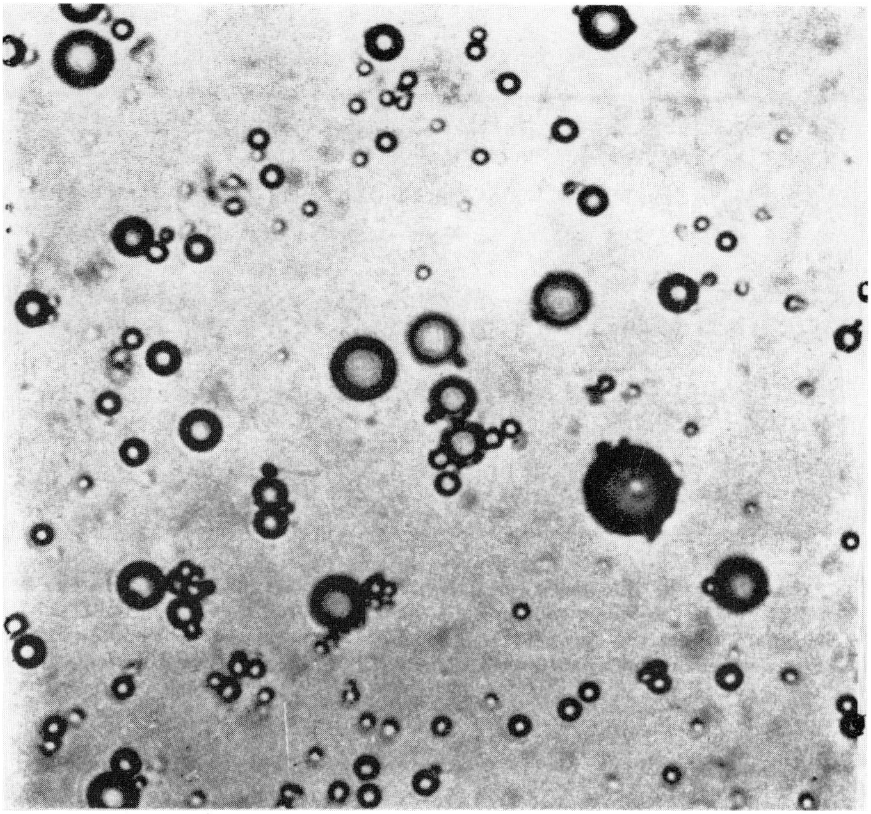

Fig. 14 (a). Light microscopic picture of **coacervates**: 0.001-1 mm in diameter (according to Oparin 1968; for details see text).

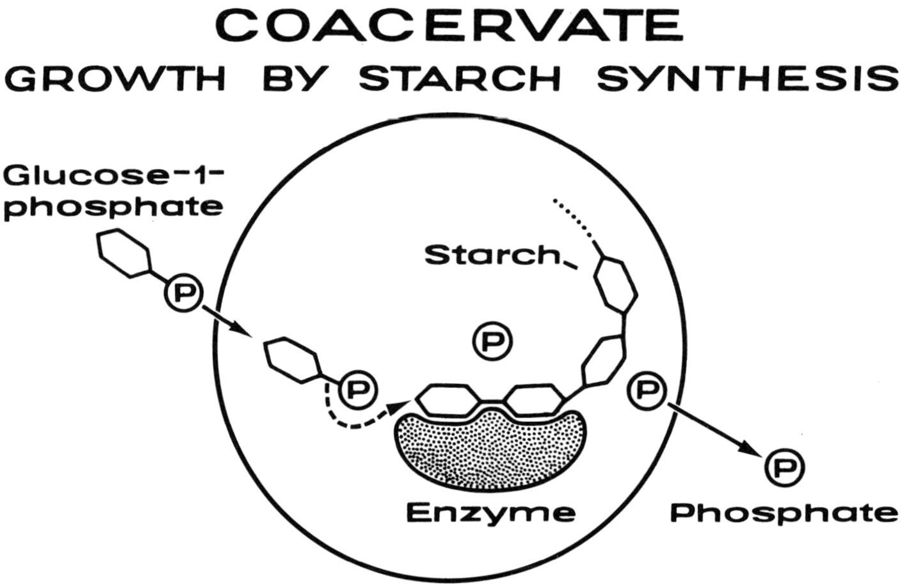

Fig. 14 (b). Schematic drawing of the growth of **coacervates** by starch synthesis (according to Dickerson 1979).

	PROPERTIES	COACERVATE	MICROSPHERE
I	DIFFERENCES:		
	MATERIAL	BIOGENETIC	PRE-BIOGENETIC
	MEMBRANE	NONE	DOUBLE MEMBRANE
	BEHAVIOUR	UNSTABLE	- STABLE - CAPABLE OF DIVISION, GROWTH, MOVEMENT
II	SIMILARITIES:		
	FORM	SPHERICAL	SPHERICAL
	DIAMETER	1 - 0.001 mm	0.08 - 0.0005 mm
	MODEL	SYSTEM OF CELL COMPONENTS	SYSTEM OF CELL COMPONENTS

Fig. 14 (c). Comparison of the significant characteristics of **coacervates** and **microspheres** (Schwemmler 1985).

treatment with supersonic vibrations, these can be dispersed to form spherical bodies called **liposomes** (e.g. Sitte 1969, Nöll 1976; Fig. 17). Prebiogenetic fatty acids, which may have been abundant on primitive earth, should also have been capable of forming such membranes.

This general method is also used to study the independent formation of cell organelles from their components in vitro. It has been shown, for example, that ribosomes, which consist of more than 50 different macromolecules, are able to organize themselves spontaneously into functional units in vitro (Garret and Wittmann 1971). Mutants of the **Escherichia coli** phage Qß in a cell-free system are even able to replicate matrix-free (and thereby form mutants), if the necessary nucleotides and linking enzymes (RNA replicases) are provided (Spiegelmann et al. 1967, Bieberich et al. 1986). In these cases, we are dealing with **evolution experiments in vitro**. They demonstrate convincingly that the independent, prebiogenetic organization of membranes and of replicative systems might in principle have been possible before the formation of a genetic apparatus capable of encoding information.

D. Computer test

The ideas on ur-cell evolution derived from experimental data have also been tested by computer simulation and probability calculations. The results of these tests of the hypercycle are so positive that **Eigen** and **Schuster**(1978) practically exclude all alternatives (Fig. 18). The nature and extent of this book, however, do not allow a presentation of the various mathematical derivations, especially since they do not permit any conclusion to be drawn concerning the presence of a membrane before the formation of the genetic apparatus, or of the genetic apparatus before the membrane. For details the interested reader is referred to the original literature (e.g. Kaplan 1978, Eigen and Schuster 1978, Kuhn 1972, 1976).

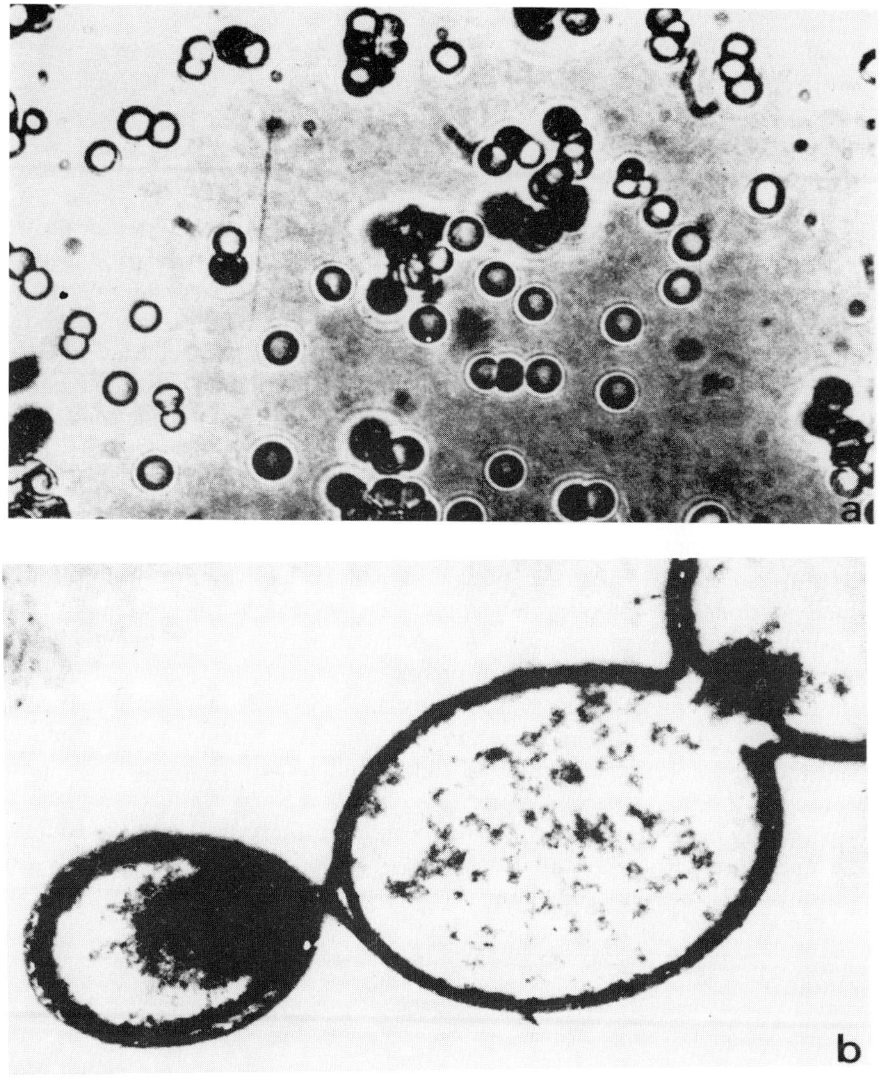

Fig. 15 (a). Light microscopic photograph of budding **microspheres**: 0.0005 - 0.08 mm in diameter (according to Fox 1965; for details see text).

(b). Electron micrograph of a cross section of a budding **microsphere** with an enveloping membrane, similar to a double membrane (according to Fox 1965).

IV. Reconstruction

A. The problem

The evidence from the various studies suggests a very early autocatalytic interaction between protein and nucleic acid as a precursor of a genetic apparatus, approximately as suggested by Eigen (hypercycle). Such an interaction seems almost to be a necessary precondition for the evolution to the ur-cell. However, it must probably also be assumed that a separate development of mixed macromolecules, up to the stage of pre-compartments, preceded and/or occurred in parallel to this "tandem evolution" of proteins and nucleic acids. Thus it is possible that precursors of membranes (e.g. prebio-membranes of fatty acids or protenoids), of metabolic processes (e.g. primitive sugar cleavage) and of the genetic apparatus (e.g. pre-ribosomes; cf. Barbieri 1981) evolved in different areas. Their development required milieus which differed from one another, in some cases so greatly that they were mutually exclusive. For example, the assumed polycondensation of proteins on nucleic acids occurs preferentially in an acid medium, while sugar cleavage occurs much more readily in an alkaline medium. In addition, the beginning of ur-cell evolution required the existence of a **semipermeable membrane** of some sort. Only such a membrane could guarantee the emergence and preservation of the reaction-specific interior milieus required for the evolution of the ur-cell. Apart from that, enclosing membranes are necessary to keep products of cleavage and synthesis in place. In addition it may be assumed that in ur-cell evolution there was an early coupling of the energy-requiring polycondensations of proteins on nucleic acids (ur-anabolism) to the energy-producing processes of sugar cleavage (ur-catabolism).

Fig. 16. Hypothetical reaction mechanism of the **simultaneous polycondensation of peptides and nucleotides** (modified, according to Krampitz et al. 1969, Paecht-Horowitz et al. 1970). In the first step, free nucleotides (NT) would couple with amino acids. Next, the nucleotide portion of the amino acyl nucleotide would add to the free 3'-end of an existing polynucleotide chain. In the third step, the amino acid portion of the dimer would couple to the polypeptide chain attached to the polynucleotide. In a fourth step, the last peptide-nucleotide bond would dissolve (see lower part of the drawing).

Simultaneous synthesis of peptides on nucleotides

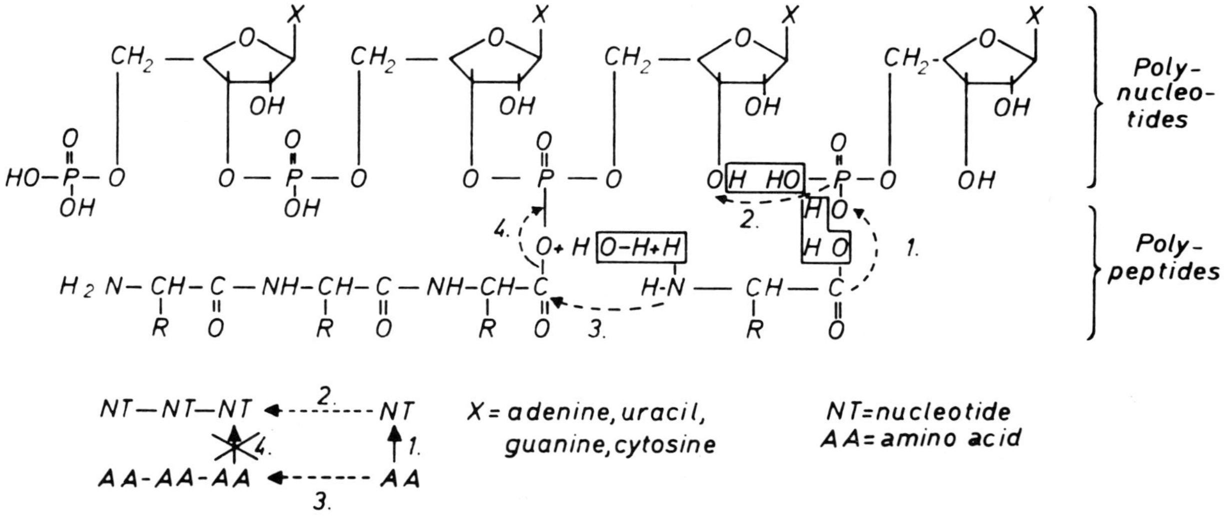

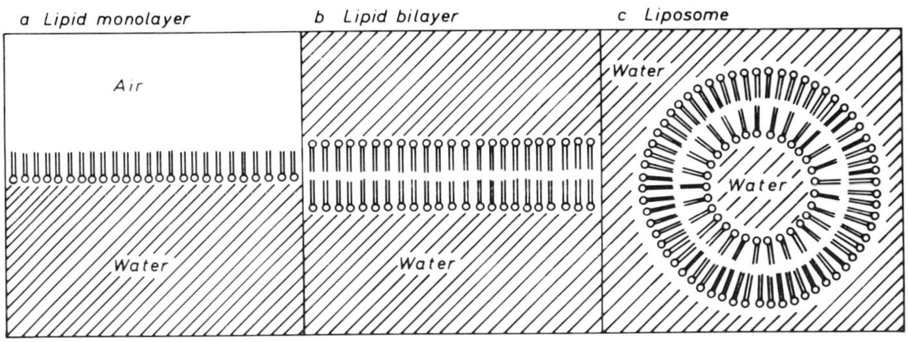

Fig. 17. **Model membrane** of amphipathic molecules such as lipids with a water-soluble and an insoluble portion (modified according to Nöll 1976; for details see text).
(a). Formation of monolayers.
(b). Formation of bilayers.
(c). Formation of liposomes, surrounded by bilayers.

Fig. 18. Hypothetical representation of ur-cell evolution according to the **hypercycle model** (adapted from Eigen and Schuster 1978). GC = Guanine/Cytosine, R = Purines, Y = Pyrimidines, N = Purines or Pyrimidines.
It may be objected that Eigen proposed his ur-cycle before it was realized that RNA can be catalytically active, e.g. in self-splicing of pre-mRNA, snRNPs, etc. The most widely-held opinion now is that the primitive RNAs were able to replicate themselves. Why and how they started directing protein synthesis remains a mystery. Polycondensation of amino acids on RNA, occurring to some degree as an accidental by-product, could be one explanation.

SCHEME	PROPERTY
A 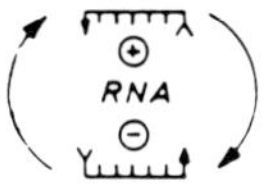	„Primitive replication"? Quasi - species (plus - and minus strands), GC - rich.
B 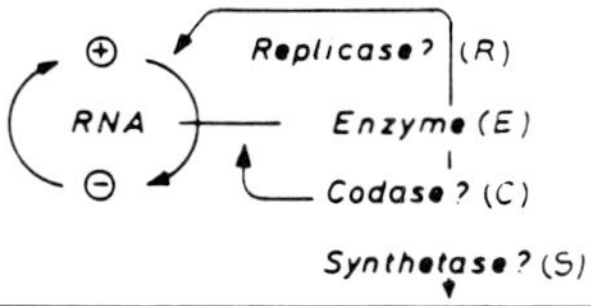	„Primitive translation" Codon assignments (plus strands? = ur-messenger, minus strands = ur-transfers), translation products (primitive enzymes?)
C 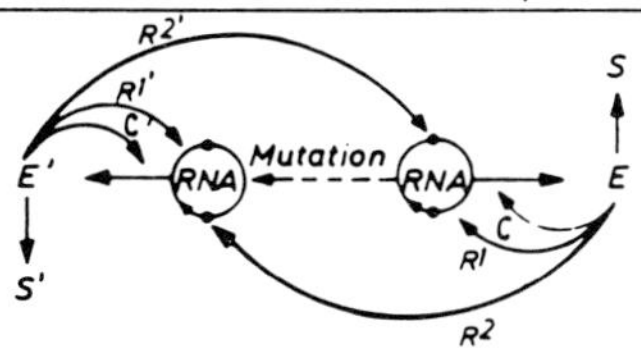	„Primitive hypercycle" Fixation of a GC-frame code through mutation and selection, assignments of glycine, alanine, aspartic acid and valine, primitive replicases, possibly with multiple function.
D 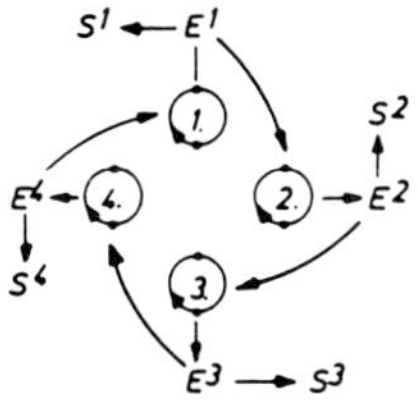	„Primitive metabolism" Evolution of hypercyclic organization, RNY code, replicases, codases, synthetases for metabolism, compartmentation (membrane?)
E 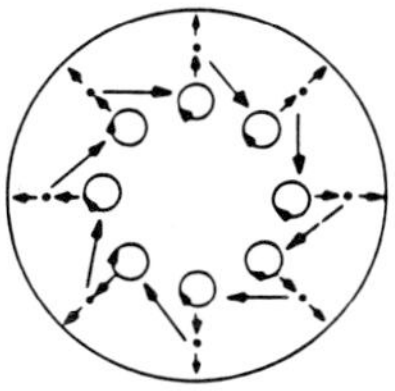	„Primitive cell" (Protobiont ?) Hypercycle enclosed in a membrane, adapted to replication and translation enzymes, development of metabolic and control functions, operon structure, RNA length comparable to viral RNA.

The postulated presence of a prebio-membrane at the beginning of ur-cell evolution and the "assembly" of the ur-cell from separate preformed pre-compartments (the precursors of protoplasm and genetic apparatus) is thus compatible, in principle, with the multi-hit hypothesis.

Seen in this light, none of the models for ur-cell evolution discussed so far meets the requirement derived above that there was a very early interaction among all three minimum components of an ur-cell: membrane, protoplasm and translation apparatus. These systems can at best serve as models for precursors of individual ur-cell compartments: **Oparin's coacervates** for the protoplasmic precursors, **Fox's microspheres** (i.a.) for membrane precursors and **Eigen's hypercycle** for precursors of a genetic apparatus. Therefore, a modified reconstruction of ur-cell evolution has been proposed which will be presented here in outline (for details, see Schwemmler 1985). It is roughly based on the multi-hit hypothesis, supplemented by parts of the multi-step hypothesis when this shows to be more appropriate. The reconstruction of the evolution of a possible minimum organism will first be sketched, and then the possible development of transition organisms on the way to a hypothetical ur-bacterium (progenote) which would have been the precursor of all presently living cells.

B. Hypercycle model

1. Protobiont

The beginning of ur-cell evolution might have been the prebiogenetic formation of double-layered membranes which, because of the alkaline sea water, would asymmetrically incorporate various prebiogenetic protenoids, leading to negatively charged upper surfaces and positively charged lower surfaces.

Fig. 19. Representation of the hypothetical asymmetric prebio-membranes formed by varying orientation of negatively or positively charged amphipathic lipids and proteins. Two possible types of liposomes are the result: **Plasma-prebioid** and **nucleo-prebioid** (according to Schwemmler 1984, for details see text).

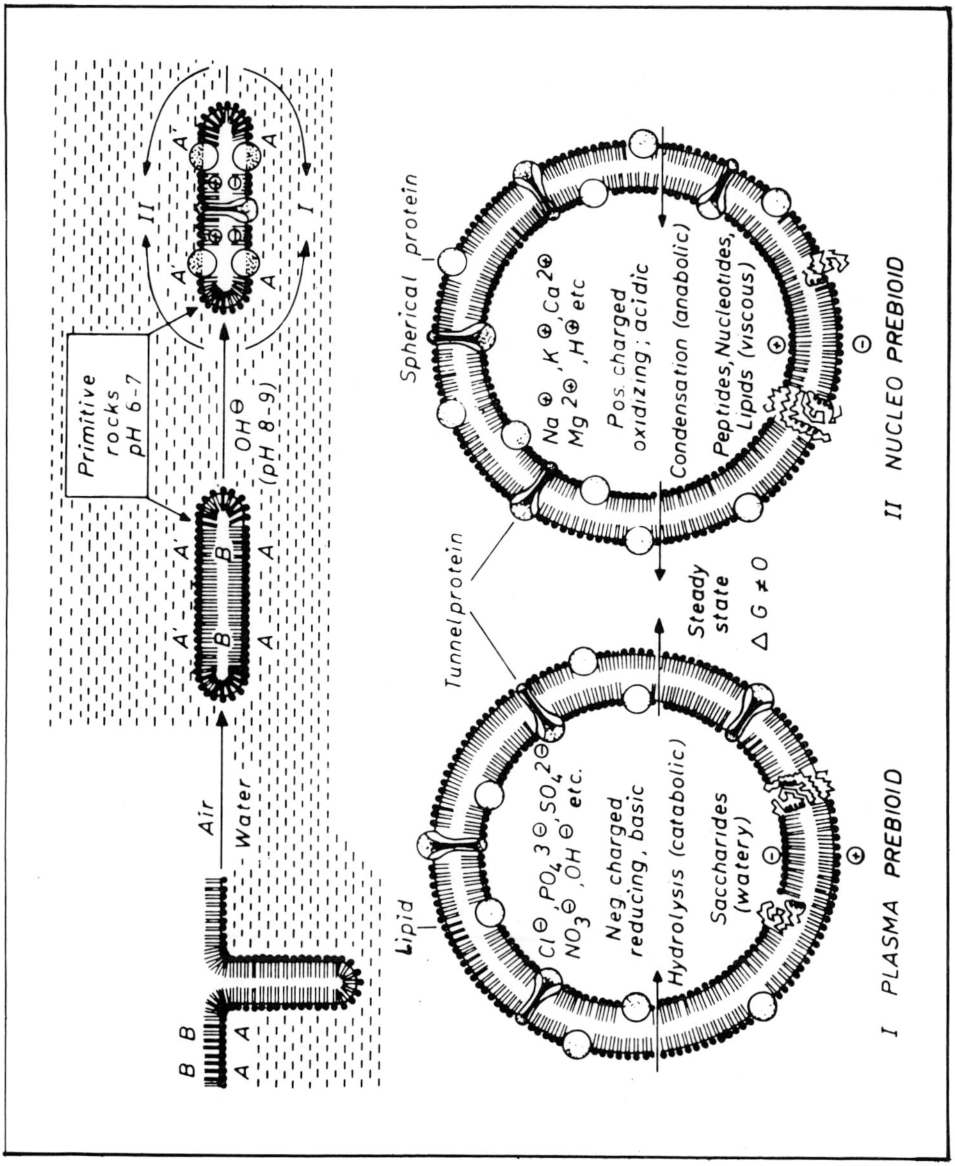
B B
A A
Air
Water
Primitive rocks pH 6-7
OH⊖ (pH 8-9)
A'
B
A
II
I
Lipid
Tunnel protein
Spherical protein
Cl⊖, PO4 3⊖, SO4 2⊖, NO3⊖, OH⊖ etc.
Neg. charged reducing, basic
Hydrolysis (catabolic)
Saccharides (watery)
Steady state
Δ G ≠ 0
Na⊕, K⊕, Ca2⊕, Mg2⊕, H⊕ etc
Pos. charged oxidizing; acidic
Condensation (anabolic)
Peptides, Nucleotides, Lipids (viscous)
I PLASMA PREBIOID
II NUCLEO PREBIOID

Dispersion of those bilayers could have led to the formation of two types of membrane-bound droplets of prebioids. Type I prebioid, with the positive side of the membrane on the outside, would have preferentially absorbed anions, including OH^-, while type II prebioids, with the negative side of the membrane outside, would have attracted and possibly taken up cations, e.g. H^+. In the first case, the interior milieu would have been alkaline, and thus reducing, while the second type of prebioid would have had an acidic or oxidizing interior (Fig. 19). In this way, through the catalytic effects of, for example, SH, OH and histidinyl groups of protenoids, the beginnings of a catabolism would have developed in the anion-rich, reducing and strongly basic type I prebioid. The selectively enriched sugars present within this prebioid, for example, would have been hydrolysed. The sugar cleavage could be seen as the beginning of a glycolytic energy metabolism, and for this reason, type I has been called the **plasma-prebioid** (Schwemmler 1984). The beginning of anabolism could develop in the cation-rich, oxidizing, slightly acidic prebioid type II through the effects of catalytically active "organic residues" of the interior membrane surface; this would have led to the simultaneous polycondensation of peptides and nucleotides. In this way, the hypercyclic coupling of proteins and nucleic acids could have started, and for this reason type II has been called the **nucleo-prebioid** (Schwemmler 1984).

Sooner or later, random collision of plasma- and nucleo-prebioids would have produced a minimum organism, the protobiont, in which the energy-providing catabolism was coupled to the energy-consuming anabolism, and genetic control was provided by development of a coding mechanism (Fig. 20). Steric conditions might have led to pairing of triplets, but probably only the middle base determined the type of the amino acid, so that in this case one may speak of a "**simplet code**" (Fig. 21a). There probably is a specific binding affinity of each of the four different nucleotide bases to one of the four different categories of amino acid side chains: basic, acidic, hydrophobic and intermediate (Fig. 21b). Since the three-dimensional structure and function of a protein depend on the

---

Fig. 20. Reconstruction of the hypothetical **minimum organism** (Schwemmler 1985).

(a) Protobionts represent the merging of energy-producing catabolic processes of the plasma-prebioid with the energy-consuming, anabolic, simultaneous synthesis of the nucleo-prebioid.

(b) Gene expression might have occurred by way of enzyme-free RNA replication and a so-called "simplet code" (for datails see text).

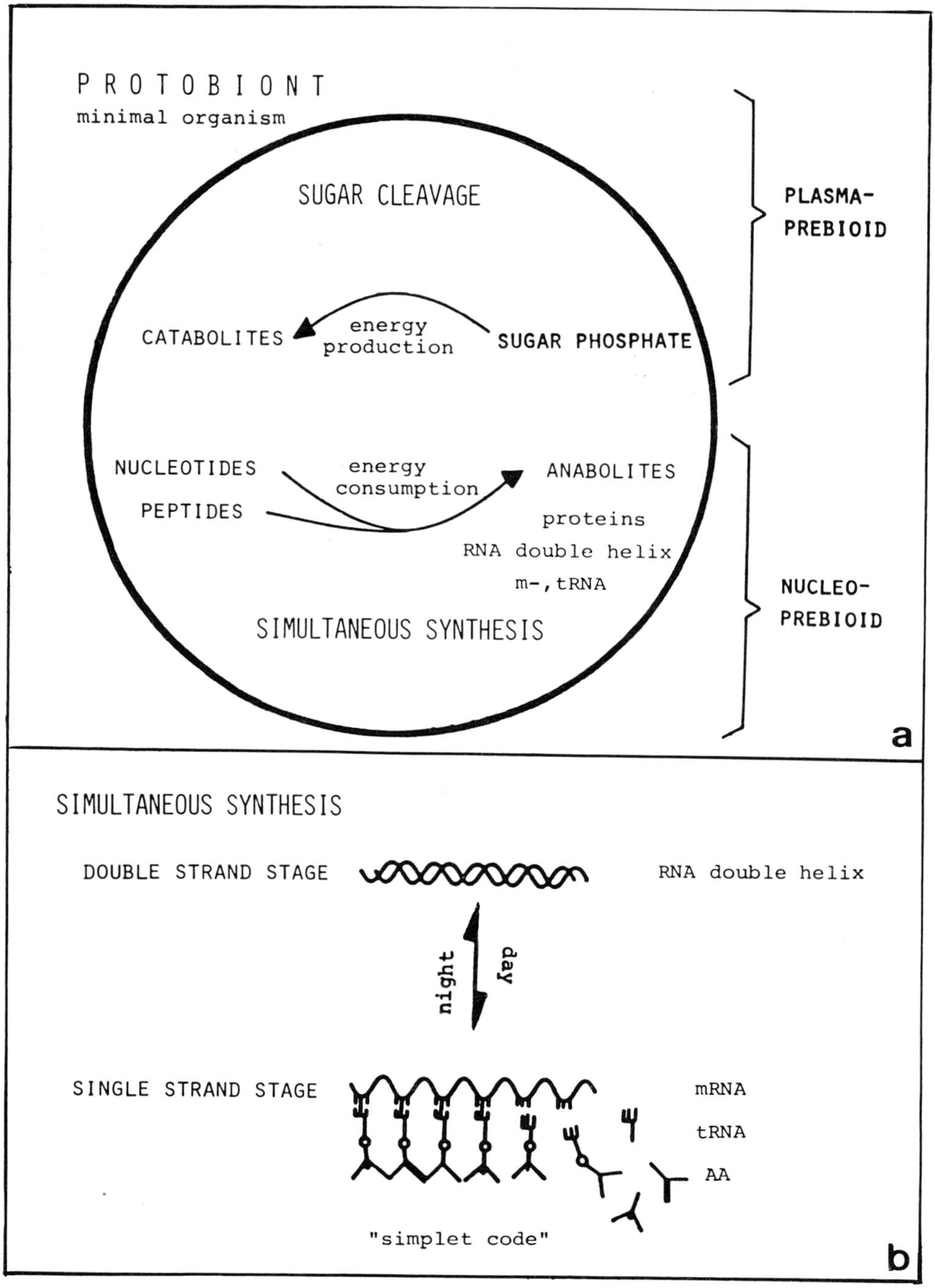
PROTOBIONT
minimal organism
SUGAR CLEAVAGE
PLASMA-PREBIOID
CATABOLITES
energy production
SUGAR PHOSPHATE
NUCLEOTIDES
PEPTIDES
energy consumption
ANABOLITES
proteins
RNA double helix
m-, tRNA
NUCLEO-PREBIOID
SIMULTANEOUS SYNTHESIS
a
SIMULTANEOUS SYNTHESIS
DOUBLE STRAND STAGE
RNA double helix
night
day
SINGLE STRAND STAGE
mRNA
tRNA
AA
"simplet code"
b

A Modern genetic code

Codon	AA	Codon	AA	Codon	AA	Codon	AA
UUy UUz	Phe Leu	UCx	Ser	UAy UAz	Tyr Stop	UGy UGA UGG	Cys Stop Trp
CUx		CCx	Pro	CAy CAz	His Gln	CGx	Arg
AUy AUA AUG	} Ile Met	ACx	Thr	AAy AAz	Asn Lys	AGy AGz	Ser Arg
GUx	Val	GCx	Ala	GAy GAz	Asp Glu	GGx	Gly

B Hypothetical original code (simplet code)

xUx(group I)	xCx(group II)	xAx(groupIII)	xGx(group IV)
Phe, Leu Ile, Val	Ser, Pro Thr, Ala	Glu, Asp	Arg Lys (?), Gly(?)
hydrophobic AA	intermediate AA	acid AA	basic AA

x^* : U, C, A or G

y^* : U or C

z^* : A or G

Fig. 21. (a) The development of the present triplet code by way of a hypothetical "simplet" or "doublet" code, whereby the middle base or the first two bases determined the type of the amino acid (according to Kaplan 1978).

* according to Eigen and Schuster (1978; cp. Fig 18); x = N; y = Y; z = R.

Fig. 21. (b) Characterization of the four different **amino acid groups** and their possible selectivity to one of the four nucleo-bases (adapted from Schwemmler 1984).

AA: amino acid; AMP: adenosine monophosphate.

AA groups / Characteristics	Charged acidic side chains: $-COO^-$	Uncharged polar side chains: $-SH$, $-NH_2$, $-OH$	Nonpolar side chains: -R	Basic side chains: $-NH_3^+$, $=NH_2^+$, $=NH^+$
H_2O solubility	Hydrophilic	Intermediate	Hydrophobic	Hydrophilic
Charge	Negative	Neutral	Neutral	Positive
Binding	Ionic	H-bridges	Hydrophobic	Ionic
Abiogenetic AA	Aspartic acid (Asp) Glutamic acid (Glu)	Asparagine (Asn) Cysteine (Cys) Serine (Ser) Threonine (Thr) Tyrosine (Tyr)	Glycine (Gly) Isoleucine (Ile) Leucine (Leu) Phenylalanine (Phe) Valine (Val) Proline (Pro) Alanine (Ala)	Arginine (Arg) Lysine (Lys)
Presently only known as biogenetic AA		Glutamine (Gln)	Tryptophan (Trp) Methionine (Met)	Histidine (His)
Rel. binding frequency in copolymerisation of AA - AMP (from Calvin 1969)	e.g. Asp - Asp 55% Asp - His 8%	e.g. Ala - Ala 47% Ala - Gly 15% Ala - Asp 2%	e.g. Val - Val 52% Val - Ala 13%	e.g. His - His 44% His - Aps 12%
Selective base affinity	Guanine ?	Cytosine	Uracil	Adenine ?

sequence of hydrophobic, acidic, basic and intermediate amino acids, it is possible that ur-enzymes of modest specificity formed with the aid of the genetic system of the protobiont.

2. *Eobiont*

It is assumed that the protobiont lacked a triplet coding for precise initiation of protein synthesis. This would have lead, due to frame shifts, to so-called "**nonsense proteins**", resulting in a loss of ability to reproduce. However, to form such an initiation system, long-chain RNA molecules would be required, and these could not evolve in the hydrolysing interior of the protobiont (Kuhn 1972, 1976). By combination of the protobiont with another **ribosome-like nucleo-prebioid** which, in a parallel development, would have selected for long-chain RNA molecules in its relatively hydrolysis-stable interior, the eobiont with a precisely functioning ribosomal protein synthesis could have arisen (Fig. 22a; cf. Barbieri 1981). The original genetic code would have developed further into a "**doublet code**"[1] in which triplets determined only with the first and middle base the type of amino acid (Fig. 22b). This would have led to a great increase in the specificity of enzymes.

---

[1] see Fig. 21 (a).

Fig. 22. Representation of the hypothetical **eobiont** (according to Schwemmler 1985).

(a) The eobiont may have developed from the combination of the protobiont with a ribosomal nucleo-prebioid.

(b) Gene expression may have been carried out by enzymatic RNA-replication, by ribosomal protein synthesis, and by a so-called "doublet code".

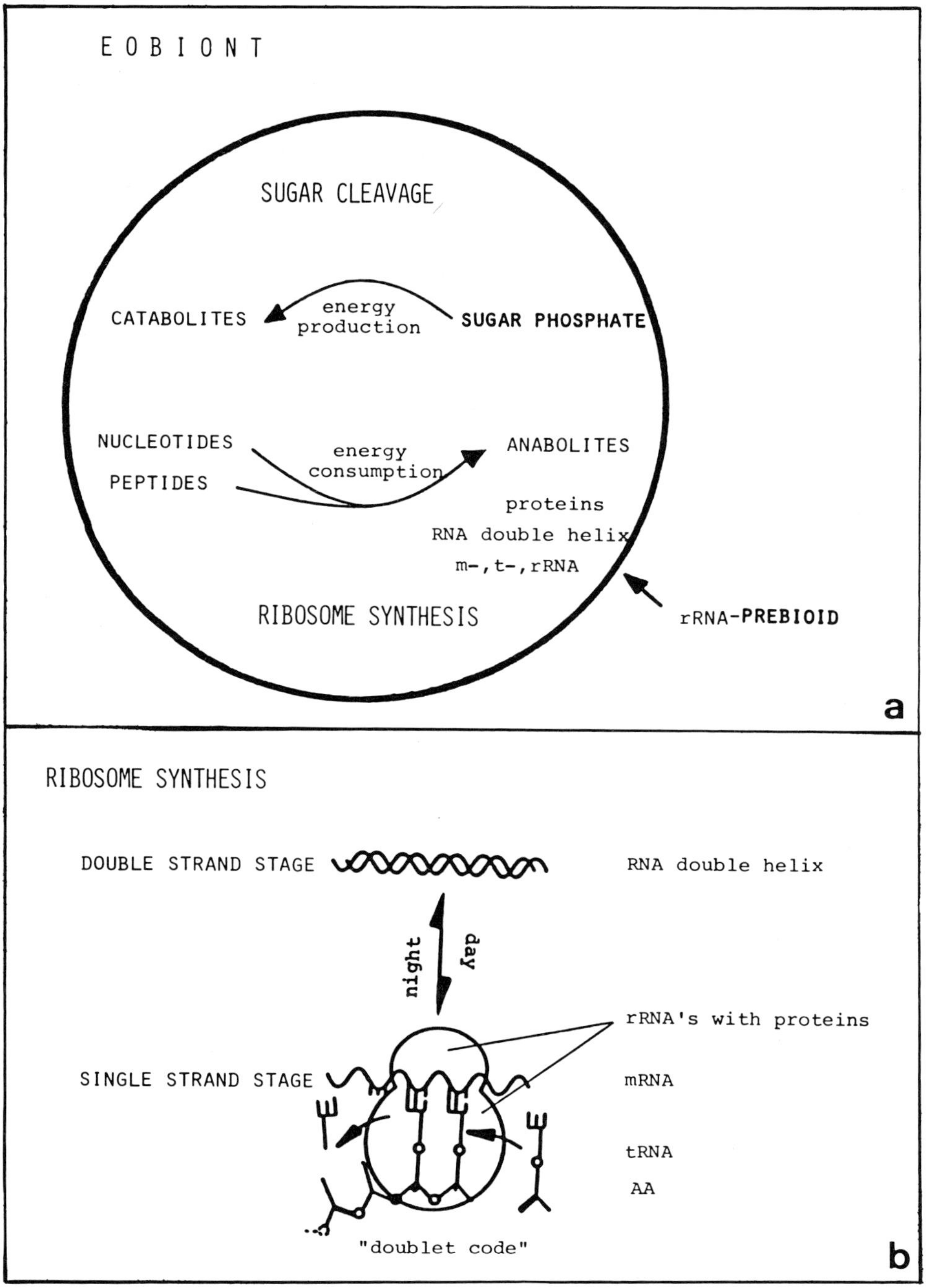

E O B I O N T
SUGAR CLEAVAGE
CATABOLITES
energy production
SUGAR PHOSPHATE
NUCLEOTIDES
PEPTIDES
energy consumption
ANABOLITES
proteins
RNA double helix
m-, t-, rRNA
RIBOSOME SYNTHESIS
rRNA-PREBIOID
a
RIBOSOME SYNTHESIS
DOUBLE STRAND STAGE
RNA double helix
night
day
rRNA's with proteins
SINGLE STRAND STAGE
mRNA
tRNA
AA
"doublet code"
b

3. Progenote

For further evolution of the eobiont, the **complementary strands** formed during replication of its diverse types of RNA - rRNA, tRNA and mRNA - supposedly were futile and interfering. As far as protein synthesis was concerned, they were useless waste products which interfered with translation of the primary strands. What is more, the primary strands were not even linked in a single information strand. Their condensation to longer RNA chains, however, was limited by the possibility of forming side chains if the 2'-OH of the ribose as well as the 3'-OH was the site of linkage. In addition, 2'-5'linkage in place of the normal 3'-5'linkage would interfere with normal helix formation. Since they would lack cooperativity, such strands would not have been able to replicate completely, and essential information would have been lost. It is thought that the genetic information could not be gathered in a single strand until **DNA** (deoxyribonucleic acid) became available which lacks the OH group in the 2'position, and therefore cannot form side chains. This development could have been initiated by combination of a **DNA nucleo-prebioid**, which had developed separately, with the eobiont, forming the ur-procyte (or progenote), a precursor of the modern bacterial cell (Fig. 23a). Here all three bases were required in the form of the **triplet code** to specify a certain amino acid; this would have made possible the synthesis of highly specific enzymes (Fig. 23b).

Fig. 23. Representation of the hypothetical **ur-procyte** or **progenote** (adapted from Schwemmler 1985).

(a) The ur-procyte may have arisen from the combination of an eobiont with a DNA nucleo-prebioid.

(a) Gene expression had probably already been carried out by RNA-transcription and ribosomal translation with a triplet code; enzymatic DNA replication had probably already evolved.

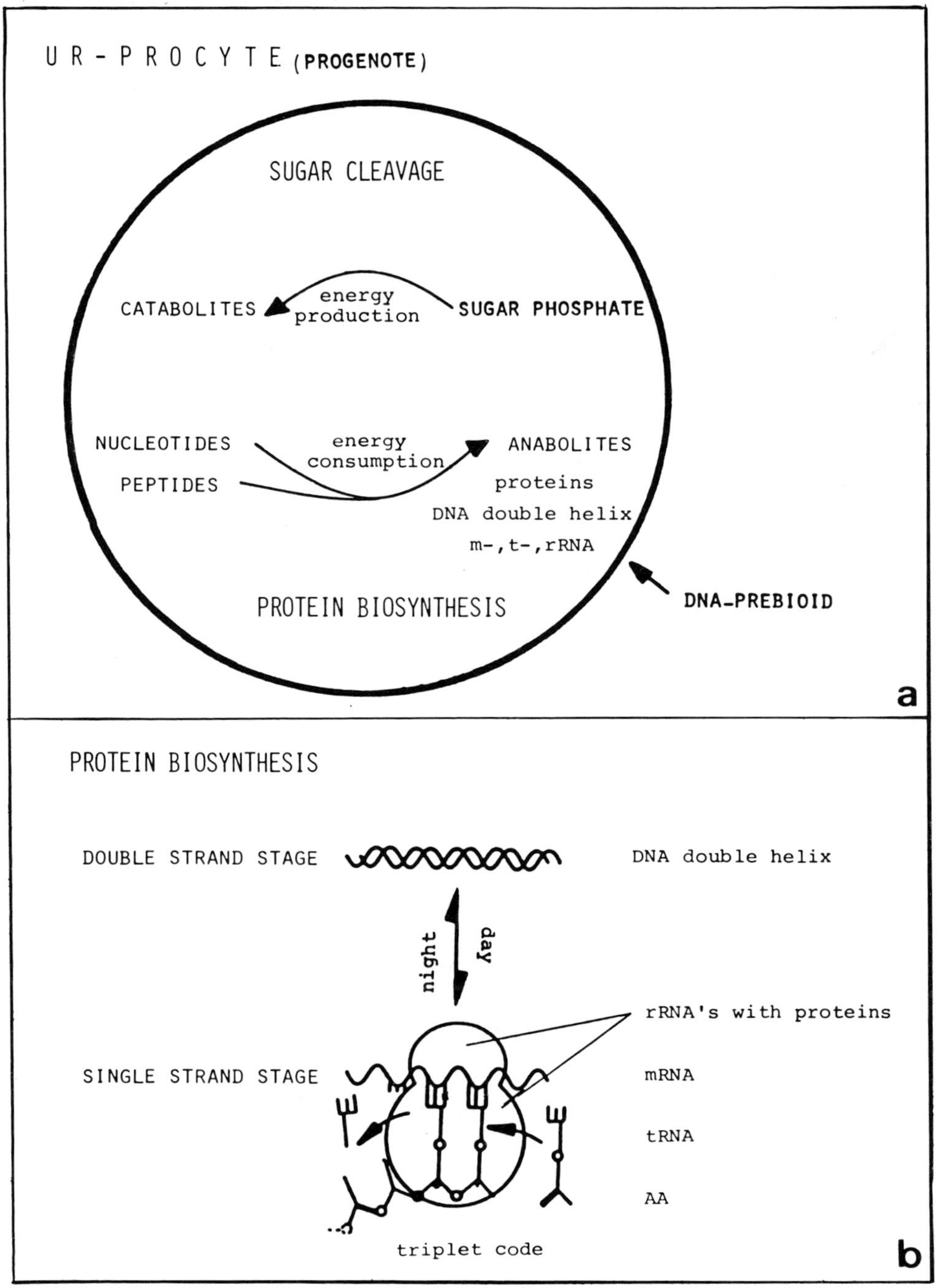
UR-PROCYTE (PROGENOTE)
SUGAR CLEAVAGE
CATABOLITES
energy production
SUGAR PHOSPHATE
NUCLEOTIDES
PEPTIDES
energy consumption
ANABOLITES
proteins
DNA double helix
m-, t-, rRNA
PROTEIN BIOSYNTHESIS
DNA-PREBIOID
a
PROTEIN BIOSYNTHESIS
DOUBLE STRAND STAGE
DNA double helix
night
day
rRNA's with proteins
SINGLE STRAND STAGE
mRNA
tRNA
AA
triplet code
b

V. Summary

The data and speculations on possible transitions from molecule to ur-cell can be summarized in six stages (Fig. 24):

1. Formation of our planet with an atmosphere of gases which could serve as raw materials for living structures.

2. Prebiotic synthesis of simple molecules (prebio-monomers) appropriate to serve biological purposes later on, such as amino acids, carboxylic acids, glycerol, sugars and nucleotide bases.

3. Combination of these molecules to the complicated macromolecules (prebio-polymers) like proteins, lipids, polysaccharides and nucleic acids (formation of the latter not proven yet).

4. Assembly of these macromolecules to various mixed aggregates which served as precursors of membranes (structure), protoplasm (metabolism) and genetic apparatus (heredity).

5. Integration of such membranes, plasma and genetic aggregates into the functioning system of a minimum organism.

6. Further development of the minimum organism via intermediate stages to a ur-procyte (progenote), the ancestral form of all presently living pro- and eukaryotic cells.

Fig. 24. Summary of the supposed transition **from molecules to the ur-procyte or progenote** (also see Fig. 9). According to this diagramm, three separate lines of evolution, rather than a single step, are proposed. The development of mRNAs and tRNAs into rRNAs could not have taken place in the same hydrolytically active environment (see text). The integration of DNA into the primitive RNA genetic system must have occurred later with the formation of long DNA molecules (see text). In addition, primitive DNA macromolecules could only develop in a hydrolytically stable inner milieu (for details of this argument see Schwemmler 1984).

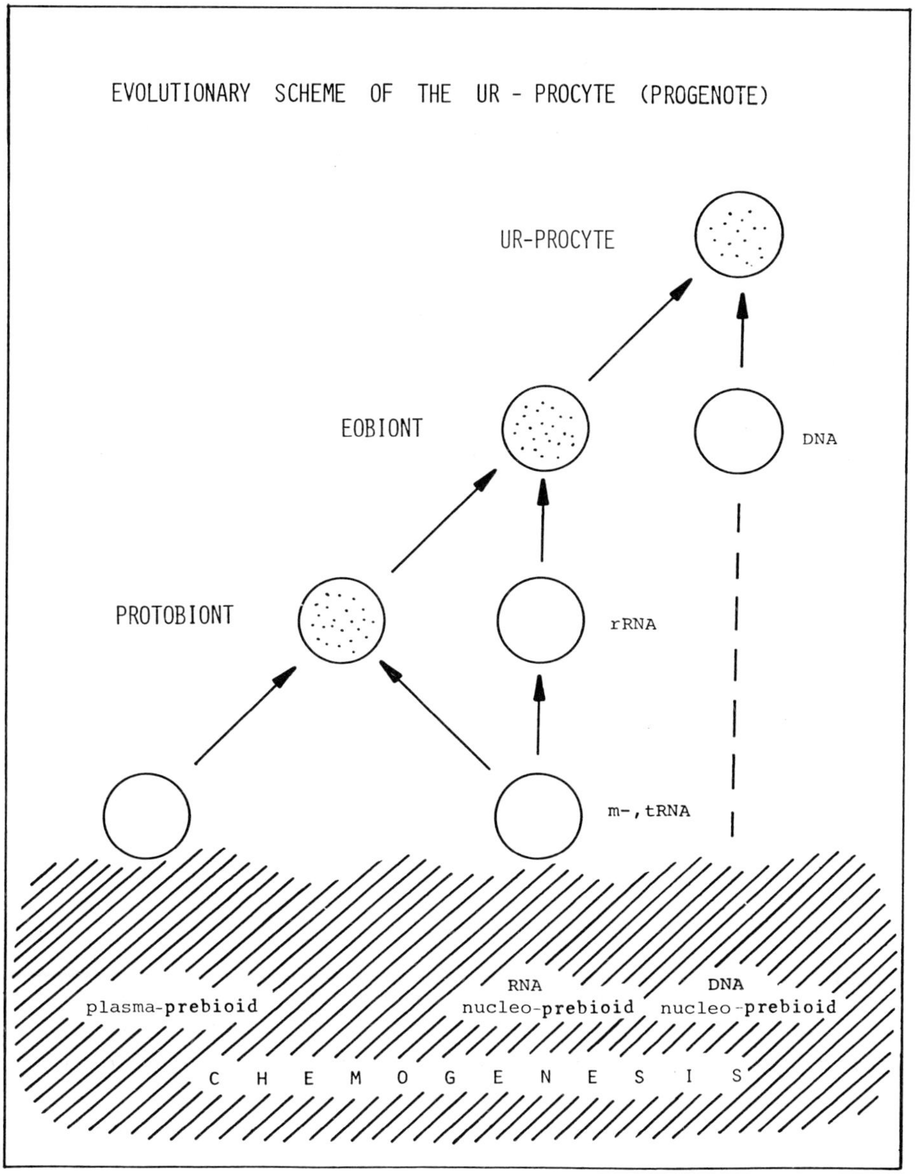
EVOLUTIONARY SCHEME OF THE UR - PROCYTE (PROGENOTE)
UR-PROCYTE
EOBIONT
DNA
PROTOBIONT
rRNA
m-, tRNA
plasma-prebioid
RNA
nucleo-prebioid
DNA
nucleo-prebioid
C H E M O G E N E S I S

VI. References

Barbieri, M. (1981): *The ribotype theory on the origin of life*. J. Theoret. Biol. 91 545-601.

Biebricher, C.K., Eigen, M. and Luce, R. (1986): *Template-free RNA synthesis by Qβ replicase*. Nature 321 89-91.

Cairns-Smith, A.G. (1982): *Genetic takeover and the mineral origins of life*. Cambridge University Press New York.

Calvin, M. (1969): *Chemical evolution*. Claredon Press Oxford.

Crick, F. (1983): *Das Leben kam vom anderen Stern*. Bild der Wissenschaft 4 118-127.

Dickerson, R.E. (1979): *Chemische Evolution und der Ursprung des Lebens*. Spektrum der Wissenschaft 9 99-115.

Dose, K., and Rauchfuß, H. (1975): *Chemische Evolution und der Ursprung lebender Systeme*. Wiss. Verlagsgesellschaft Stuttgart.

Eigen, M. (1971): *Self-organization of matter and the evolution of biological macromolecules*. Naturwissenschaften 58 465-523.

Eigen, M. and Schuster, P. (1978): *The hypercycle. A principle of natural self-organization. Part C: The realistic hypercycle.* Naturwissenschaften 65(7) 341-369.

Fox, S.W. (1965): *The origin of prebiological systems and their molecular matrices.* Academic Press New York.

Fox, S.W. and Dose, K. (1972): *Molecular evolution and the origin of life*. Freeman, San Francisco.

Garret, R.A. and Wittmann, H.G. (1971): unpublished data.

Hoyle, F. and Wickramasinghe, C. (1982): *Wie das Leben auf die Erde kam. Eine neue Evolutionstheorie wird diskutiert*. Bild der Wissenschaft 1 39-48.

Kaplan, R.W. (1978): *Der Ursprung des Lebens. Biogenetik, ein Forschungsgebiet heutiger Naturwissenschaft*. 2nd edition; Georg Thieme Verlag Stuttgart.

\- (1982): *Probleme des Lebensursprungs: Wie waren die ersten lebenden Systeme beschaffen?* Forum Mikrobiologie 5 230-237.

Krampitz, G., Baars, S., Haas, W. and Kempfle, M. (1969): *Zur Kondensation von Aminoacyladenylaten. Ein Modell für abiogene Proteinsynthese.* Naturwissenschaften 56 416.

Kuhn, H. (1972): *Selbstorganisation molekularer Systeme und die Evolution des genetischen Apparates*. Angew. Chemie 84 (18) 838-860.

\- (1976): *Model consideration for the origin of life*. Naturwissenschaften 63 68-80.

Miller, S.L. (1953): *A production of amino acids under possible primitive earth conditions*. Science 117 528-529.

Monod, J. (1970): *Le hazard et la nécessité*. Editions du Seuil, Paris.

Nagy, B. (1976): *Organic chemistry on the young earth.* Naturwissenschaften 63 499-505.

Nöll, G. (1976): *Modellmembranen - Membranmodelle.* Biologie in unserer Zeit 6(3) 65-74.

Oparin, A.I. (1924): *Origin of life.* Moscow.

\- (1968): *Genesis and evolutionary development of life.* Academic Press New York.

"Origin of life" (1986): Abstracts of the 5th ISSOL Meeting. Berkely ISSOL.

Paecht-Horowitz, M., Berger, J. and Katchalsky, A. (1970): *Prebiotic synthesis of polypeptides by heterogenous polycondensation of amino acid adenylates.* Nature 228 636-639.

Pflug, H.D., Meinel, W., Neumann, K.H. and Meinel, M. (1969): *Entwicklungstendenzen des frühen Lebens auf der Erde.* Naturwissenschaften 56 10-14.

Schwemmler, W. (1984): *Reconstruction of cell evolution: A periodic system.* CRC Press Boca Raton (Florida).

\- (1985): *Possible Transitions from Molecules to Cells.* J. Theor. Biol. 117 187-208.

Sitte, P. (1969): *Biomembran: Struktur und Funktion.* Ber. Dtsch. Bot. Ges. 82 329-383.

Spiegelmann, S., Mille, D.R. and Petersen, R.L. (1967): *An extracellular Darwinian experiment with a self-duplicating nucleic acid molecule.* Proc. Natl. Acad. Sci. USA. 59 217-224.

Second Subchapter Procyte

I. Phenomenon

Bacteria in general, including the **cyanobacteria**, which were formerly called blue-green algae, make up the kingdom of **prokaryotes**. The cells of these organisms, the so-called **pro(to)cytes**, do not have true, membrane-bounded nuclei like those of the eukaryotes; instead they have only nuclear equivalents (Fig. 25). These characteristically consist of a double-helical DNA molecule closed upon itself to form a ring, embedded in the protoplasm and bound at certain points to the cell membrane. The procyte protoplasm contains a few structures (e.g. flagellum, mesosome, thylakoids, reserve substances), but otherwise shows little intracellular compartmentalisation. In addition, procytes are almost exclusively unicellular, and are potentially immortal, since they reproduce by transverse cell division, and thus each cell continues its existence in its progeny. Only in exceptional cases is there a formation of multicellular systems with tissue specialization, culminating in dead cell systems (e.g. the multicellular fruiting bodies of slime bacteria or myxobacteria). The almost general unicellularity and short life spans, along with the general lack of intracellular compartmentation and true speciation (inability to crossbreed) led to an absence of pronounced individual or phylogenetic development among the procytes, which does not apply to eucytes. Morphogenesis is thus of little use as a means for studying phylogenetic relationships among procytes.

Comparisons according to the **criteria of homology** developed by **Remane** (1952) are therefore more successfully applied to procytes at the level of **metabolic physiology** and **biochemistry**. An important first result of such considerations is the derivation of universal basic metabolic types, such as **fermentation**, **short-chain respiration** (including sulfur- and methane- auto-trophism), **full-chain respiration** (including chemo-autotrophism), **photergism**[1] and **photosynthesis**. Each of these types occurs in both the anaerobic (oxygen-avoiding) and the aerobic (oxygen-requiring) form, and is listed here in the order of increasingly effective energy production. In the process of fermentation, energy is obtained by cleavage of sugar phosphates in the protoplasm. In photergism, energy is

---

[1] **Photergism** is defined by **Kaplan** (1978) as ATP production (without NAD(P)H production) by light-driven processes. In this special case, the term means ATP synthesis by a protein-bound carotene derivate (rhodopsin) and sunlight.

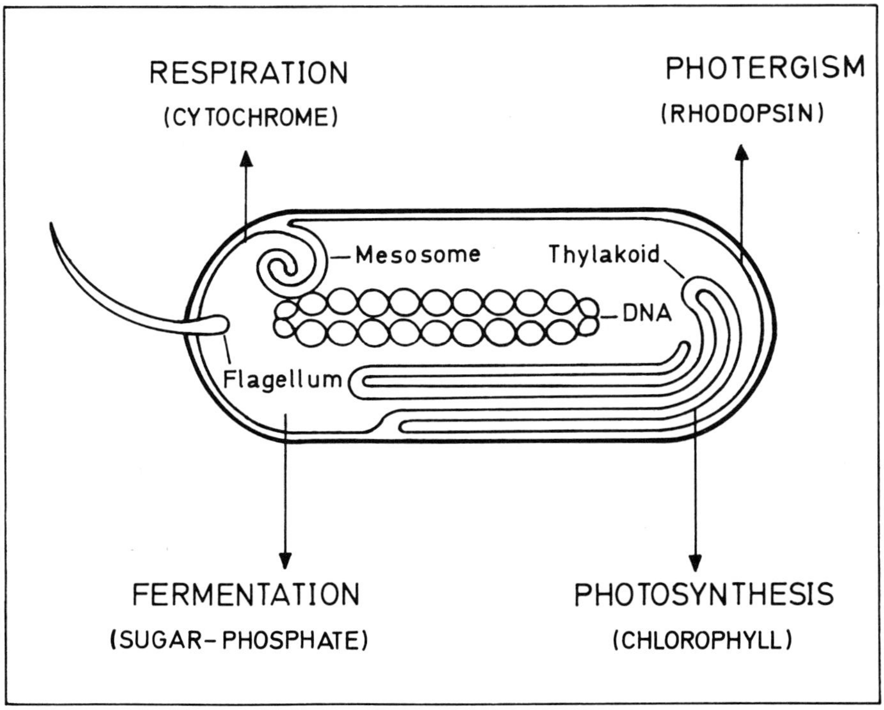

Fig. 25. Combined scheme for some types of procyte metabolism (Schwemmler 1985).

Fig. 26. Characteristics of important metabolic types of procytes.

	ENERGY	BIOTOPE	REPRESENTATIVES
FERMENTATION	cleavage: sugar phosphates	dark, anaerobic	some acidophilic archaebacteria, clostridia
PHOTERGISM	proton pump: i.a. light + rhodopsin	light, (an)aerobic	halobacteria
PHOTOSYNTHESIS	redox systems: i.a. light + chlorophyll	light, (an)aerobic	photobacteria, cyanobacteria (bluegreen bacteria)
SHORT-CHAIN RESPIRATION (e.g. sulfur-autotrophism: without Calvin cycle)	redox systems: i.a. cytochrome *c*	dark, (an)aerobic	sulfur-dependent archaebacteria (e.g. *Desulfurococcus*), eubacteria (e.g. *Desulfovibrio*)
LONG-CHAIN RESPIRATION (including chemo-autotrophism: with Calvin cycle)	redox systems: i.a. cytochrome oxidase	dark, (an)aerobic	intestinal bacteria, soil bacteria

obtained through a light-driven proton pump associated with the membrane pigment rhodopsin. In photosynthesis, energy is obtained from light-driven, membrane-bound chlorophyll, and in respiratory metabolism, including sulfur-, methane- and chemo-autotrophism[1], from a membrane-anchored respiratory chain consisting of a series of coupled redox systems (e.g. the cytochromes; Fig 26). There are essentially two hypotheses concerning the order in which these metabolic types evolved.

In discussing them, we shall not be concerned with the evolution of individual metabolic processes of procytes, especially since in that case a number of other variants would also have to be introduced. Furthermore, a detailed presentation would have to take into account the fact that many basic metabolic types have arisen in different ways. There are procytes with basically different photosynthesis apparatus (e.g. **Heliobacterium chlorum**: P I[2]: **Chloroflexus**: P I; **Rhodobacteria**: P I; **Chlorobium**: P II[3]; **Cyanobacteria**: P I and P II). Thus, founded on the present state of our knowledge, a general classification of basic types of metabolic pathways will be given, beginning with a discussion of their evolutionary derivation, just as has been done with the basic morphological categories in plants and animals.

II. Hypotheses

According to the older **conversion hypothesis** (e.g. Broda 1975), the fermenters were at the beginning of procyte evolution. The photosynthesizers were thought to have arisen from the fermenters. Finally the respirers were believed to have developed, by loss or "conversion" of the photosynthetic apparatus, and the halobacteria (photergic organisms) from the respirers. The more recent **splitting hypothesis** (review of literature see Margulis 1981, Schwemmler 1984), in contrast, suggests that the anaerobic fermenters first gave rise to anaerobic short-chain respirers, which in turn divided into two branches: the photergic and the

---

[1] In sulfur- and methane-autotrophism, CO_2 fixation appears to occur exclusively via a reductive tricarboxylic acid cycle, whereas the other chemo-autotrophic bacteria use the Calvin cycle for CO_2 fixation.

[2] P I : cyclic photosynthetic system;
[3] P II : acyclic photosynthetic system.

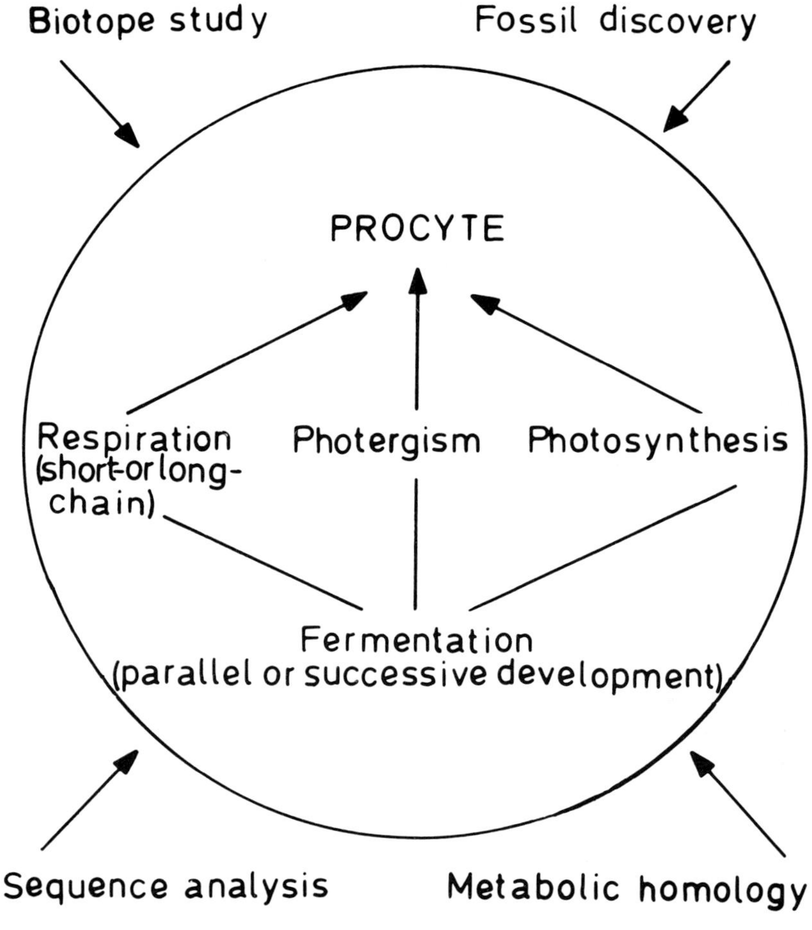

Fig. 27. Alternative hypotheses and methods of analysis in procyte evolution.

photosynthetic organisms, which later gave rise to the long-chain respirers. Both branches would have led, via anaerobic intermediates, to aerobic forms. The essential difference between the two hypotheses is the time at which the chlorophyll system has supposedly arisen. According to the conversion hypothesis, the chlorophyll system evolved before the cytochrome system, whereas the splitting hypothesis postulates that the chlorophyll system could not have arisen until at least the first parts of the cytochrome system were present.

III. Data

The available data will now be reviewed, to see which hypothesis they may confirm (Fig. 27).

A. Biotope study

Geology and **mineralogy** have provided valuable clues to the evolution of procytes. It has been found that the atoms of the oldest marine rocks are mainly reduced, i.e. in their lowest valence state. In younger marine rocks, e.g. in the banded iron ores, starting about 2.8×10^9 years ago, the oxidized forms of the elements begin to dominate. It is therefore assumed that the iron dissolved in the primitive ocean was originally present as divalent Fe^{2+}, but that it was gradually oxidized to Fe^{3+} by the oxygen which began to appear about 3×10^9 years ago, and was deposited by periodic episodes of precipitation in "banded" form, as mixed $Fe^{2+,3+}O_4$. In this way, oxygen is thought to have been kept out of the atmosphere for a time. Its production is ascribed to the activity of photosynthesizing microorganisms which, as fossil evidence indicates, by that time were present in the seas in sufficient quantity .

After iron and other oxidizable elements in the ocean had been exhausted, oxygen accumulated in the atmosphere. In the areas of contact between atmosphere and lithosphere, the oxygen then converted the terrestrial iron to $Fe_2^{3+}O_3$, and the red, fully oxidized terrestrial ores were deposited. The last "banded" marine ores are about 1.8×10^9 years old, which is also the approximate age of the first oxidized iron ores on land.

Thus, the ecological interpretation of the geological evidence indirectly indicates the existence of chlorophyll systems more than 3×10^9 years ago, but does not permit conclusions as to whether the cytochrome system arose before that time or afterwards. It would therefore seem worthwhile to examine the data from microfossils to see whether they can contribute to an answer to this question.

B. Fossil discovery

The task of paleochemistry is to examine ancient rock layers for **chemical fossils**, using sensitive methods like gas chromatography, which enable conclusions to be drawn concerning biotic systems (for a review, see Dose and Rauchfuß 1975). In doing so, the subsequent formation of decay products from original compounds is taken into account. For example, the presence of porphyrin in the $3.4x10^9$ - $3.2x10^9$ year old Onverwacht Layer in South Africa might be a decay product of cytochrome or of chlorophyll, since both of these contain porphyrin rings. The occurrence of isoprenoids in the $3.2x10^9$ year old Fig Tree formation of South Africa, by contrast, can only indicate the presence of chlorophyll, which contains such isoprenoids in its side chain. Chlorophyll, in turn, is an indication of photosynthetic activities. The enrichment of the carbon isotope ^{12}C with respect to ^{13}C also indicates the selective activity of photoassimilating microorganisms. However, a later diffusion of such organic molecules from younger to older rock layers cannot reliably be excluded. Therefore, there is much uncertainty involved in the interpretation of chemical fossils.

More reliable witnesses are the huge limestone reefs called stromatoliths, which are more than $3x10^9$ years old. These are believed to be mineralized bacterial colonies. Today, mineralized bacterial colonies of this type are produced chiefly by photosynthetic bacteria; reefs produced this way are so similar to geologically ancient formations that the latter are considered unequivocal evidence for bacterial photosynthesis. **Microfossil** analyses of thin slices of ancient sediments also serve as indicators. The Fig Tree layer ($3.2x10^9$ - $3.1x10^9$ years old) contains fossils which are almost certainly remains of microorganisms (including **Eobacterium isolatum**). Morphological and paleochemical analyses suggest that they are either anaerobically respiring (?) sulfur bacteria or (an)aerobically respiring (?) photergic halobacteria (for a review, see Dose and Rauchfuß 1975; Kaplan 1978). In younger Fig Tree formations ($2.7x10^9$ years old) a microorganism has been found (**Archaeosphaeroides barbertonensis**) which is comparable to the green sulfur bacteria; accordingly, anaerobic photosynthesizers may have existed at least this early. This appears to be confirmed by discoveries of the same age in the Bulawayo layer in South Africa and in the Soudan formation in North America. A possible interpretation of the microfossil analyses would thus allow the inference that parts of the cytochrome system developed before bacterial chlorophyll. However, since paleological interpretation of simple impressions of tiny spheres or rods is always extremely uncertain, we shall draw upon comparisons of basic metabolic processes in procytes to substantiate such an interpretation.

C. Metabolic homology

In addition to the universal genetic code, procytes with various types of metabolism are all capable of comparable basic biochemical processes. They all contain the same assortment of 150 different molecules, including amino acids, nucleotides, sugars, fatty acids and so on, as well as the macromolecules of proteins, nucleic acids, polysaccharides and lipids. The biosynthesis (anabolism) of these basic units is essentially the same, with **ATP** as the "general energy currency" and **NADH** as the "key substance" of hydrogen transport. These and other criteria suggest that all procytes are phylogenetically related (Schön 1978).

There are differences between procyte metabolic types with respect to their sources of energy and carbon, and their metabolic substrates. Biosynthesis in the fermenters, photergers and respirers depends on organic carbon, i.e. on the existence of other organisms or of prebio-polymers (compare p. 26; Fig. 28). This condition is generally called **"carbon (C-) heterotrophism"**. Most photosynthesizers, sulfur-, methane- and chemo-autotrophs, however, use carbon dioxide as their only source of carbon, and are thus independent of other organisms. Therefore this phenomenon is called **"C-autotrophism"**.

The energy for anabolism and for preservation of the complex cell structures is provided by energy metabolism. There are two basic forms: in **chemotrophic** energy metabolism (catabolism), chemical energy released by degradation of organic substrates is conserved in the form of ATP or GTP (fermentation, respiration; sulfur-, methane- or chemo-autotrophism). In **phototrophic** metabolism, energy is obtained by conversion of light energy into chemical energy, which in turn is bound as ATP or GTP (photosynthesis, photoergism).

Chemotrophic and phototrophic energy metabolism can be further subdivided. In chemotrophic metabolism, energy is obtained by oxidation (dehydrogenation), in

---

Fig. 28. Overview of **procyte metabolism** (for details, see text).

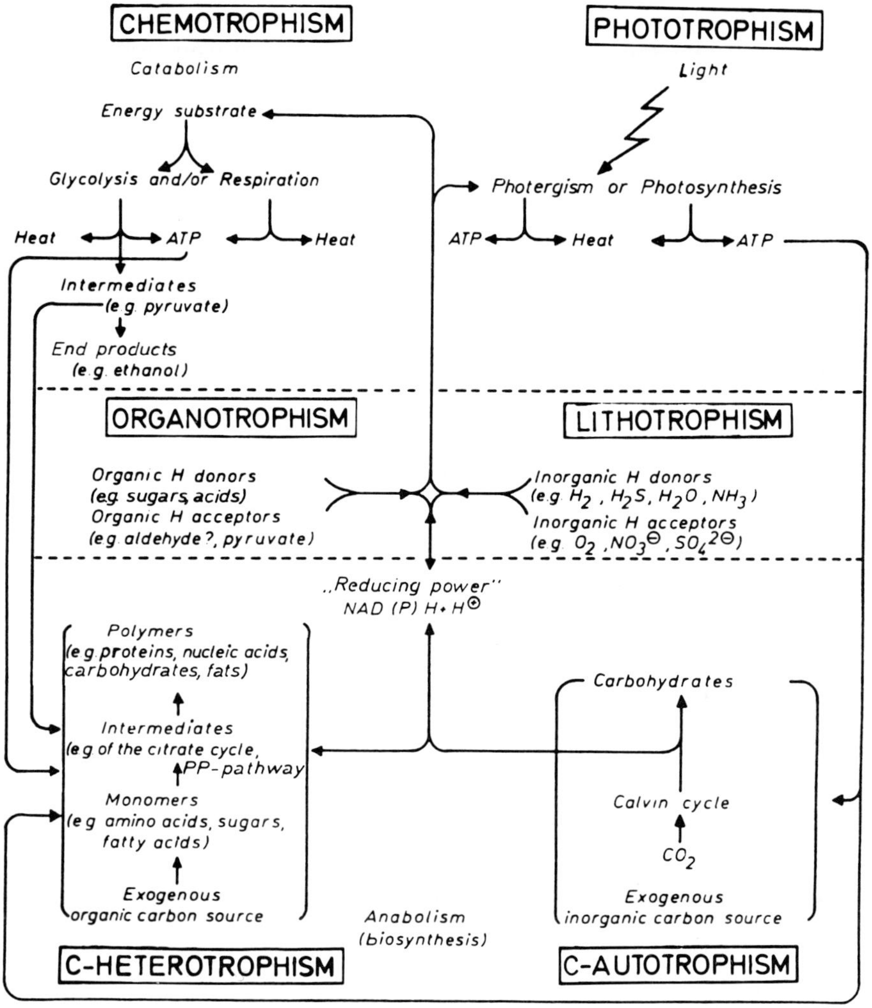
CHEMOTROPHISM
PHOTOTROPHISM
Catabolism
Light
Energy substrate
Glycolysis and/or Respiration
Photergism or Photosynthesis
Heat
ATP
Heat
ATP
Heat
ATP
Intermediates
(e.g. pyruvate)
End products
(e.g. ethanol)
ORGANOTROPHISM
LITHOTROPHISM
Organic H donors
(e.g. sugars, acids)
Organic H acceptors
(e.g. aldehyde?, pyruvate)
Inorganic H donors
(e.g. H_2, H_2S, H_2O, NH_3)
Inorganic H acceptors
(e.g. O_2, $NO_3^{\ominus}$, $SO_4^{2\ominus}$)
„Reducing power"
NAD (P) H + $H^{\oplus}$
Polymers
(e.g. proteins, nucleic acids, carbohydrates, fats)
Intermediates
(e.g. of the citrate cycle, PP-pathway)
Monomers
(e.g. amino acids, sugars, fatty acids)
Exogenous
organic carbon source
Carbohydrates
Calvin cycle
CO_2
Exogenous
inorganic carbon source
Anabolism
(biosynthesis)
C-HETEROTROPHISM
C-AUTOTROPHISM

which hydrogen is transferred from the substrate (H-donor) to a hydrogen acceptor (H-acceptor, e.g. NAD(P)). In the simple case, the H-donor and H-acceptor are both **organic** compounds, as in fermentation. In respiration, usually only the H-donors are organic, while the H-acceptors are, almost without exception, **inorganic**. In sulfur-, methane- and chemo-autosynthesis, both **H-donors and H-acceptors** are used. Thus fermentation and respiration are chemo-organotrophic types of metabolism, while sulfur-, methane- and chemo-autosynthesis are chemo-inorganotrophic (=chemo-lithotrophic) types. In phototrophic energy metabolism, too, there are both organic and inorganic H-donors, which in this case supply hydrogen to the two most important cellular H-acceptors, NAD and NADP. The reduced forms of these compounds represent a "reducing power" which is used, among other things, for reduction of CO_2 to carbohydrates (photosynthesis: Calvin cycle). One can distinguish between an (an)aerobic photo-organotrophic metabolism (photergism, primitive type of photosynthesis) and an aerobic photo-inorganotrophic (=photo-lithotrophic) metabolism (more highly developed type of photosynthesis).

On the basis of their energy metabolic types - additional metabolic characteristics are not discussed here (Margulis 1981, Schwemmler 1984) - a hypothetical evolutionary tree of basic metabolic types of procytes may be constructed (Fig. 29). According to this reconstruction, the basic types of procytes would have evolved in the order shown below:

Fermenters

--> Anaerobic respirers (sulfur-, methane-autotrophism)

--> Photergers

--> Photosynthesizers

--> Aerobic respirers (chemo-autotrophism)

Comparison of metabolic types thus appears to confirm unequivocally, although indirectly, that at least the first part of the cytochrome system arose before the appearance of bacterial chlorophylls. It is therefore only reasonable to determine to what degree sequence analysis can provide direct evidence for this hypothesis.

Fig. 29. Probable **phylogenetic tree of procytic metabolic types**, based on comparison of metabolism and additions of new redox systems.

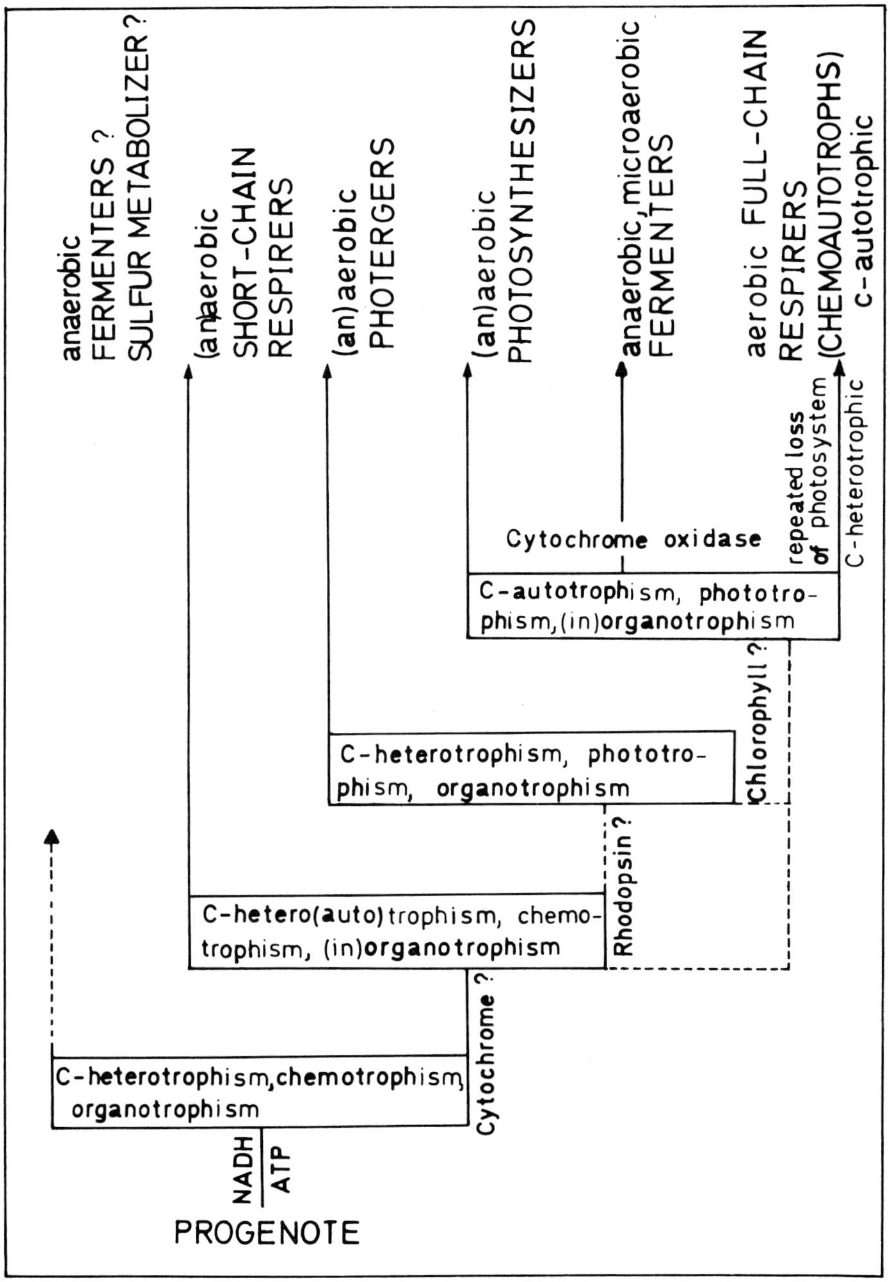
PROGENOTE
NADH
ATP
C-heterotrophism, chemotrophism, organotrophism
Cytochrome ?
anaerobic FERMENTERS ? SULFUR METABOLIZER ?
C-hetero(auto)trophism, chemo-trophism, (in)organotrophism
(an)aerobic SHORT-CHAIN RESPIRERS
Rhodopsin ?
C-heterotrophism, phototro-phism, organotrophism
(an)aerobic PHOTERGERS
Chlorophyll ?
C-autotrophism, phototro-phism, (in)organotrophism
(an)aerobic PHOTOSYNTHESIZERS
Cytochrome oxidase
anaerobic, microaerobic FERMENTERS
repeated loss of photosystem
aerobic FULL-CHAIN RESPIRERS (CHEMOAUTOTROPHS)
C-heterotrophic
c-autotrophic

	L(6)	M(9)	K(8)	K*(7)	c_{555}(2)	c_{553}(1)
L(6:c_2,c_{550})	49.7	38.5	17.8	17.5	15.6	12.0
M(9:c_2,c)	38.5	51.6	18.9	19.1	15.1	11.9
K(8:c_2,c_{551})	17.8	18.9	43.8	17.6	13.8	14.4
K*(7:c_{554},f)	17.5	19.1	17.6	43.6	18.4	6.4
c_{555}(2)	15.6	15.1	13.8	18.4	47.0	12.0
c_{553}(1)	12.0	11.9	14.4	6.4	12.0	-

Fig. 30. Identity matrix for cytochrome c of DNA-containing cell-organelles from various procytes, established by analysis of amino acid sequence between homologous cytochromes of the most important type classes (according to Dickerson 1980; for details, see text).

The identity matrix is a standard for the phylogenetic relationship of different species.

Fig. 31. The essential aspects of the **phylogeny of procytic** metabolic types can be derived from the sequences and structures of **cytochromes c**, various nucleic acid molecules and all other available data (adapted from Dickerson 1980). According to these data, parts of the cytochrome system developed before the appearance of chlorophylls. Aerobic respirers, however, are thought to have arisen from phototrophic bacteria which had lost their ability to photosynthesize (e.g. **Pseudomonas, Paracoccus, Beggiatoa, Leucothrix**). Mitochondria appear to have arisen from the non-sulfur purple bacteria, and chloroplasts from the cyanobacteria (see p. 118); the vertical axis of the scheme does not exactly represent the time scale.

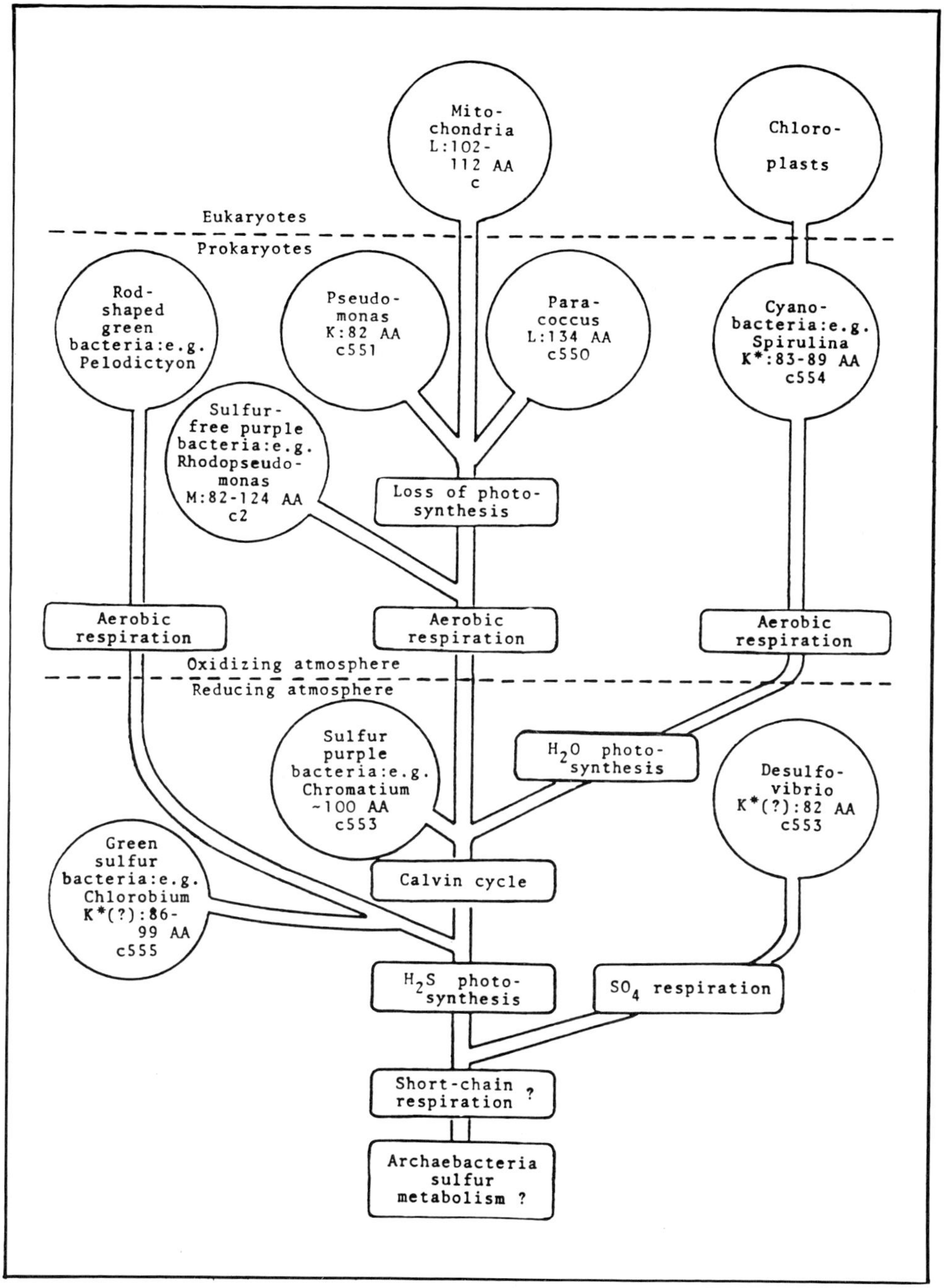

Mito-
chondria
L:102-
112 AA
c
Chloro-
plasts
Eukaryotes
Prokaryotes
Rod-
shaped
green
bacteria:e.g.
Pelodictyon
Pseudo-
monas
K:82 AA
c551
Para-
coccus
L:134 AA
c550
Cyano-
bacteria:e.g.
Spirulina
K*:83-89 AA
c554
Sulfur-
free purple
bacteria:e.g.
Rhodopseudo-
monas
M:82-124 AA
c2
Loss of photo-
synthesis
Aerobic
respiration
Aerobic
respiration
Aerobic
respiration
Oxidizing atmosphere
Reducing atmosphere
Sulfur
purple
bacteria:e.g.
Chromatium
~100 AA
c553
H_2O photo-
synthesis
Desulfo-
vibrio
K*(?):82 AA
c553
Green
sulfur
bacteria:e.g.
Chlorobium
K*(?):86-
99 AA
c555
Calvin cycle
H_2S photo-
synthesis
SO_4 respiration
Short-chain
respiration ?
Archaebacteria
sulfur
metabolism ?

D. Sequence analysis

Procytes preserve a record of their phylogenetic past in the amino acid sequences of their **proteins** and in the nucleotide sequences of their **deoxyribonucleic** and **ribonucleic acids**. These living archives reach far into the past and are more reliable than fossils and deductions from comparisons of metabolism together. In order to uncover this rich treasure of direct information, it was necessary to develop methods for determining the order of nucleotides in RNA- and DNA-encoded genes, or of amino acids in the proteins produced by the genes (for details, see Woese 1981). The determination of a nucleotide sequence provides three kinds of phylogenetic information:

1) disclosure of degrees of relatedness
2) inferences on the duration of evolution
3) reconstruction of original traits.

The greater the similarity in analogous genes of two organisms, the more closely these organisms are related to one another. On the other hand, the degree of difference between the nucleotide sequences of analogous genes provides a clue to the amount of time which has passed since the two organisms began to diverge. Finally, the original version of a gene may be reconstructed from a large set of homologous sequences.

Since the information encoded in genes is generally transcribed "word for word" in the RNA produced from them, and translated directly into the proteins synthesized by the RNA, the sequences of the two types of gene product can also be used for phylogenetic consideration. Although sequence analyses of nucleic acids have only been carried out in the past few years, protein sequence data from as far back as 30 years ago are available. **Dayhoff** (1972) began storing sequence data from nucleic acids and proteins (e.g. of ATP synthetases, ferredoxins, cytochromes c) in computers and searching them systematically for homologies.

Cytochrome c, a protein involved both in respiration and photosynthesis, has proved especially suitable in phylogenetic analysis of procyte metabolic types. According to **Dickerson** (1980), cytochrome c can be seen as "the red thread of Ariadne" in the labyrinthine development of bacterial metabolism. His sequence analyses corroborate the classification of cytochromes c into four classes, L, M, K and K*, which was originally based on X-ray studies of structure. This can be shown by an identity matrix obtained by comparison of amino acid sequences of two homologous cytochromes: the number of sites at which the amino acids are identical is counted and entered in a matrix. The identity matrix given here shows average values for the cytochrome c molecules of four different size-classes (Fig. 30). Evidently the number of identical amino acids in cytochromes of the same class (rectangles) is much larger than in those of different classes. The M and L cytochromes (dotted rectangles) are an exception; they could be included in a single class, if they did not differ markedly in structure. The numbers in parentheses give the number of cytochromes of each class which was included in the study. Dickerson does not include cytochrome $\underline{c}_{555}$ (green sulfur bacteria) and $\underline{c}_{553}$ (Desulfovibrio) in the K group, because they occupy a special group on the basis of their amino acid sequences.

The **phylogeny** shown in Fig. 31 is derived from the result of structure and sequence analysis of cytochrome $\underline{c}$; it shows some essential aspects of the evolution of procytic metabolic types. It can be seen that anaerobic short-chain respiration probably developed before anaerobic photosynthesis, i.e. that parts of the cytochrome system arose before the chlorophylls. In addition, it can be inferred from the scheme that aerobic full-chain respiration (cytochrome oxidase) developed much later. It arose independently in the various lines of bacteria, which had already diverged. This is also indicated by the fact that the bacterial cytochrome $\underline{c}$ reacts easily with the corresponding cytochrome reductases from mitochondria, but is fairly unreactive with mitochondrial cytochrome oxidases. This is exactly what may be expected, assuming that the cytochrome reductase was inherited from a common ancestor of all prokaryotes, while cytochrome oxidase is polyphyletic. An alternative possibility is that the oxidase-coding genes were transferred from one bacterial strain to another in the form of suitable plasmids[1]; such **horizontal gene transfer** cannot be completely excluded.

---

[1] Plasmids are ring-shaped, extrachromosomal DNA molecules which can replicate either when integrated into bacterial DNA or independently. They confer on their recipients additional characteristics like antibiotic resistance or conjugation behaviour.

IV. Reconstruction

A. The problem

All procytes seem to be phylogenetically related according to their biochemical and genetic homologies. The vast diversity of procyte forms can finally be traced back to a few basic metabolic types, including fermentation, short- and long-chain respiration (e.g. sulfur-, methane- and chemo-autotrophism), photergism and photosynthesis. The facts and inferences discussed above indicate that the derivation of photosynthesis from anaerobic short-chain respiration is more probable than the reverse. Thus at least parts of the cytochrome system probably arose before the first bacterial chlorophylls. This essentially agrees with the divergence hypothesis. On the other hand, it can be considered proven that the aerobic full-chain respirers or chemo-autotrophs arose from aerobic photosynthesizers by loss of the photosynthetic apparatus. In these cases the corresponding parts of the cytochrome system, the cytochrome oxidases, could only have evolved after the bacterial chlorophylls, which in principle would agree with the conversion hypothesis. The transitions from one type of metabolism to another could presumably have been caused by incisive changes in the biotope and in the availability of substrates (Fig. 32).

Clearly both conversion and divergence hypotheses are partially valid, as often is the case with "alternative" hypotheses. The phylogenetic relationships between the procyte metabolic types, deduced by **Dickerson** on the basis of cytochrome sequence analyses, should be reproduced with due caution. Bacterial phylogeny based on sequence analyses of single proteins is incomplete and therefore not always reliable. For example, cytochromes of the c type do not occur in all bacteria, do not always have the same function, and are therefore not completely comparable. In addition, because the bacterial lines are older than the eucyte lines, they may have been subjected to greater divergence in amino acid sequences. Aside from the above, the phenomenon of **horizontal gene transfer** in the course of bacterial evolution could lead to occasional misinterpretations in the construction of such phylogeny. Horizontal gene transfer could have played a more important role in procyte evolution than is generally realized. Genes for the enzymes of special metabolic capacities, possibly including those for cytochrome oxidases, in the course of evolution could have been transferred from one bacterial system to the other in the form of plasmids (cf. p. 85). An example of horizontal gene transfer is provided by the gene for the red plant pigment

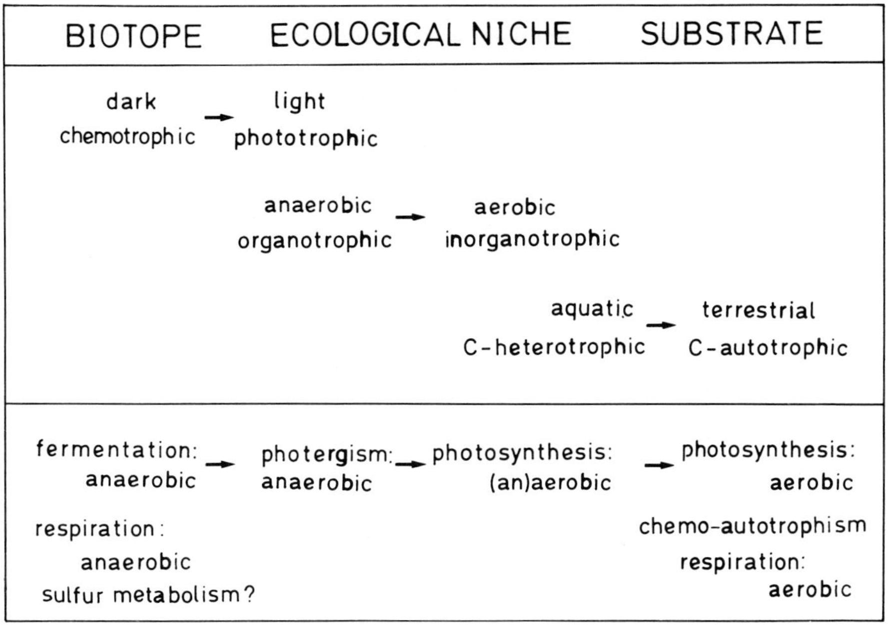

Fig. 32. Hypothetical scheme of development of the important basic types of procytes in connection with incisive changes in biotope or substrate.

leghemoglobin, activated in pulses (legumes) by their endocytobionts[1] (see Dayhoff 1972). This pigment is homologous to vertebrate hemoglobin. A direct transfer of the corresponding genes from the plant to the animal system, or conversely, is not possible. It follows that sequence homologies in this case are of exogenous origin and do not indicate phylogenetic relatedness.

For these reasons, **sequence analyses of 16S ribosomal** RNA molecules[2] (these correspond to the 18S RNA molecules in eukaryotes) **in the various types of bacteria are taken as the best basis of phylogenetic reconstruction.** Ribosomal RNAs are found universally, have conserved their function over long periods of evolution, and are very conserved in structure because they are constrained to interact with other proteins in a highly complex structure. From homologous ribosomal RNA sections, **Woese** and **Fox** (1977) infer common ancestors, i.e. phylogenetic affinity. However, taking into account the possibility of horizontal gene transfer, occasional misinterpretation cannot be excluded.

The reconstruction of procyte evolution is therefore carried out as a short, hypothetical outline based on a combination of experimental data. The most important contribution is Woese's (1981) well substantiated, widely accepted scheme for the archaebacteria (cf. Doolittle and Bonen 1981), although it must be remembered that even with this scheme, important questions remain unanswered (cp. König 1986, Stackebrandt 1986):

> *1) It has not been shown yet whether fermentation is actually the phylogenetically oldest type of metabolism, or a later "reductive" adaptation to special organic substrates.*
>
> *2) Since the oldest[3] known living bacteria are sulfur respirers, this metabolic type could with equal justification be regarded as the original one.*

Owing to these and still other unsettled questions, the reconstruction of procyte evolution offered here must be regarded as preliminary.

---

[1] Intracellular symbionts which live with their hosts in a state of mutual benefit.

[2] S stands for Svedberg unit, a measure of the rate of sedimentation in the ultracentrifuge, and thus indirectly also of the molecular weight.

[3] as shown by comparative sequence analysis.

B. Archaebacterium model

The term "**progenote**" was introduced by **Woese** and **Fox** (1977) for the presumed common ancestor of all procytes, the ur-bacterium. Progenotes would have been much more primitive than recent organisms. Since living representatives are not known, progenotes, at least the first ones, are simply extrapolations and projections of relevant data into the grey predawn of procyte evolution (compare p. 56). In the course of their development, the progenotes appear to have diverged into three main lines: the **archaebacteria**, the **eubacteria** and the **ur-karyotes**, the progenitors of the later eukaryotes (Fig. 33).

1. Archaebacterium

The term "**Archaebacterium**" was first introduced to scientific literature in 1977 by **Woese et al.**. They applied it to a group of exotic bacteria which are adapted to extreme conditions. They look like bacteria and, like all prokaryotes, lack nuclei. In their biochemistry and in the structures of certain large molecules (cell wall, membrane lipids, transcriptional and translational apparatus, enzymes, cofactors, etc.), however, they differ from other prokaryotes as greatly as they do from the eukaryotes (details see Fig. 34). Phylogenetically, they belong to neither of them. Archaebacteria form an entirely new kingdom, holding their own position in systematics and in the history of life (Kandler and Zillig 1986). In addition, most of them exhibit a kind of metabolism adapted to conditions which presumably existed on primitive earth (compare p. 22-25). According to Woese, archaebacteria are probably the oldest living group of organisms whereby it cannot be concluded that the inferred ur-karyotes must necessarily have been phylogenetically younger. Recent representatives of the archaebacteria are the methanogens, the extreme halophiles (halobacteria) and the thermoacidophile sulfur-dependents.

The **methanogens** are the most divergent of this group, and are widely distributed. They live in stagnant water including sewage treatment plants, in the stomach of ruminants and in animal digestive tracts in general, and even on the bottom of the ocean and in hot springs. They are thus found in anaerobic biotopes (oxygen kills them), where they are associated with bacteria which decompose organic material. These bacteria form the carbon dioxide and the hydrogen the methanogens require for the synthesis of their own biomass. They use hydrogen to reduce CO_2 to methane and release it as "swamp gas".

The aerobic **halobacteria** require high salt concentrations, and some of them even thrive in saturated solutions of sodium chloride which they seem to be able to turn red (Woese 1981). They are endowed with the comparably primitive photosynthetic apparatus of photergism (compare p. 71, also p. 122): bacterial rhodopsin, as a light-driven, membrane-bound proton pump, synthesizes ATP (Fig. 35). Bacterial rhodopsin is remarkably similar, though not homologous, to the pigment of the eye (Keniry et al. 1984). Halobacteria are able to maintain a large concentration gradient for certain ions across their cell membrane; using the potential energy of this gradient, they are able to transport numerous substances into and out of the cell, and to induce the motility of the flagellum.

The strictly anaerobic **thermoacidophilic** sulfur-dependents and the aerobic thermoacidophilic bacteria are also "exotics" with unusual living conditions. The thermoacidophiles prefer hot sulfur springs with temperatures of 80-90 °C and pH values lower than 2. Thermoplasmas live on smouldering coal heaps and, like the mycoplasmas, have no cell wall. Thermoplasmas will die if the temperature drops.

2. Eubacterium

Woese (1981) includes the majority of bacteria in the group of true bacteria, or **eubacteria**. Their nucleotide sequences are completely different from those of the eukaryotes and of the archaebacteria. The phylogenetic relationships among the eubacteria given by Woese et al. (1977) contradict traditional phylogeny. Woese distinguishes six main branches. The Gram-positive bacteria have a thick cell wall which can be specifically stained with a method developed by Gram, a Danish bacteriologist. The photosynthetic purple bacteria and some related forms (ancestral forms of mitochondria), which presumably arose by loss of the photosynthetic apparatus, make up another group, as do the spirochetes. The cyanobacteria are very likely the ancestors of chloroplasts (details, see p. 118). The cocci and the anaerobic, phototrophic green bacteria form the last two groups. Four of the branches of eubacteria contain photosynthetic

---

Fig. 33. Simplified hypothetical **reconstruction of procyte evolution** (for details see text; Schwemmler 1986).

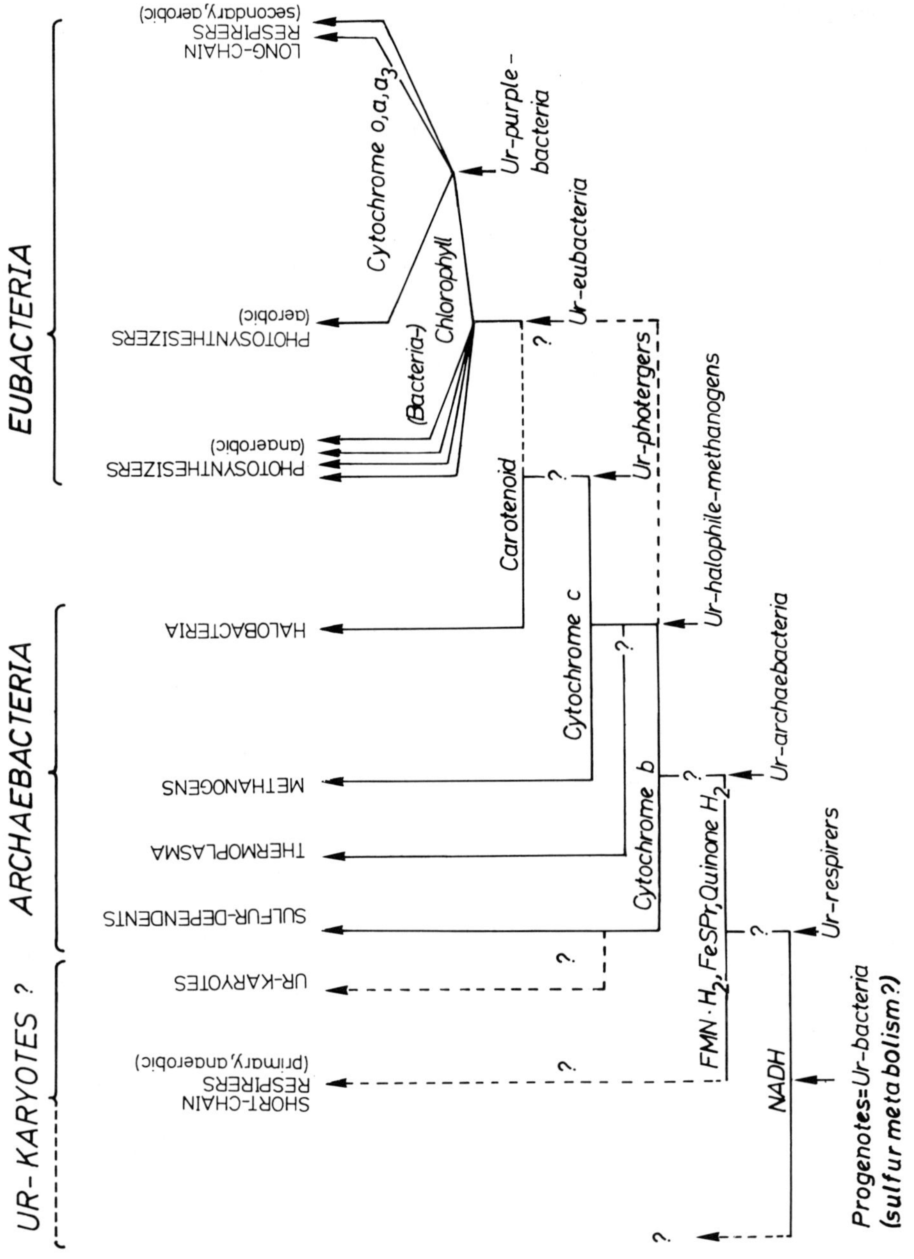
EUBACTERIA
ARCHAEBACTERIA
UR-KARYOTES ?
LONG-CHAIN RESPIRERS (secondary, aerobic)
PHOTOSYNTHESIZERS (aerobic)
PHOTOSYNTHESIZERS (anaerobic)
HALOBACTERIA
METHANOGENS
THERMOPLASMA
SULFUR-DEPENDENTS
UR-KARYOTES
SHORT-CHAIN RESPIRERS (primary, anaerobic)
Cytochrome o,a,a3
(Bacteria-) Chlorophyll
Carotenoid
Cytochrome c
Cytochrome b
FMN·H2, FeSPr, Quinone H2
NADH
Ur-purple-bacteria
Ur-eubacteria
Ur-photergers
Ur-halophile-methanogens
Ur-archaebacteria
Ur-respirers
Progenotes=Ur-bacteria (sulfur metabolism?)

representatives. From this, Woese concluded that the eubacteria have common phototrophic ancestors.

3. Ur-karyote

The **ur-karyotes** are a hypothetical line which Woese postulates to be the precursors of nucleus and cytoplasm of the eukaryotic system. He infers the existence of such an independent line from (among other things) the fact that the nucleus appears to contain at least three types of genes: eubacterial genes (presumably incorporated from DNA-containing organelles: mitochondria and chloroplasts); archaebacterial genes (e.g. the gene for a ribosomal protein), and those of unknown origin (for example, cytoplasmic ribosomal RNA). To regard the ur-karyotes as a group diverging from the archaebacteria or as direct descendants of the progenotes is evidently a matter of speculation at present.

V. Summary

The arguments presented here show that the order of the redox systems in the respiratory chain corresponds to their increasing redox potential, and at the same time to their evolutionary development, **so that the chlorophyll system cannot have evolved before the first parts of the cytochrome system** (Fig. 36). The respiratory chain can thus be used as a guide-line through the evolution of procytes (cf. also Fig. 33). The details are:

The primitive progenotes may have had a fermentation system located in their cytoplasm. Subsequent types of bacteria then successively developed in their

---

Fig. 34. The three large groups of **archaebacteria, eubacteria** and **eukaryotes** are classified on the basis of characteristic features in molecular structure. Certain features are found only in archaebacteria; others are also found in a similar form in eubacteria and eukaryotes. Thymine and dihydrouracil can replace uracil in RNA transfer molecules. The elongation factor is part of the translation apparatus of the cell (according to Woese 1981).

	Archaebacteria	Eubacteria	Eukaryotes
Cell length	Approx. 1 µm	Approx. 1 µm	Approx. 10 µm
Subcellular organelles	Absent	Absent	Present
Nuclear membrane	Absent	Absent	Present
Cell wall	Many types, none containing muramic acid	Many varieties of one type, all containing muramic acid	No cell wall in animal cell, otherwise many types
Membrane lipids	Ester with branched chains	Ester with straight chains	Ester with straight chains
Transfer RNA molecules: thymine in the same side chain	Absent	Present in most tRNAs of most species	Present in all species in most tRNAs
Dihydrouracil	Present in one genus only	Present in all species in most tRNAs	Present in all species in most tRNAs
Amino acid of the starter tRNA	Methionine	Formylmethionine	Methionine
Ribosomes: size of subunits	30S, 50S	30S, 50S	40S, 60S
Approximate length of 16S (18S) RNA	1500 nucleotides	1500 nucleotides	1800 nucleotides
Approximate length of 23S (25S, 28S) RNA	2900 nucleotides	2900 nucleotides	3500 nucleotides or more
Elongation factor	Reacts with diphtheria toxin	Does not react with diphtheria toxin	Reacts with diphtheria toxin
Sensitivity to:			
-chloramphenicol	Insensitive	Sensitive	Insensitive
-anisomycin	Sensitive	Insensitive	Sensitive
-kanamycin	Insensitive	Sensitive	Insensitive
Messenger RNA binding sites AUCACCUCC at the 3'end of 16S (18S) RNA	Present	Present	Absent

Type of retinal protein	Absorption maxima, λ max.	Amino acid sequence	Molecular weight:dalton	Function
1. Bacteriorhodop-sin	568 nm	248	ca. 26.500	Photochemical ATP synthesis
2. Halorhodopsin	578 nm	251	ca. 27.000	"
3. P 370	370 nm	?	?	Sensorial photoreceptor
4. Photorhodopsin =sR-II480=P480	480 nm	?	23.-24.000	"
5. P 565	565 nm	?	?	"
6. sR-I587	373,587 nm	?	23.-25.000	"
7. Vertebrate rhodopsin	500 nm	348	39.048	"

Fig. 35. Overview of characteristics and numbers of **retinal protein pigments** so far known of halobacteria, compared with vertebrate rhodopsin. Rhodopsin 1. and 2. convert light energy for ATP synthesis, rhodopsin 3. to 7. work as sensorial photoreceptor (-pigment).

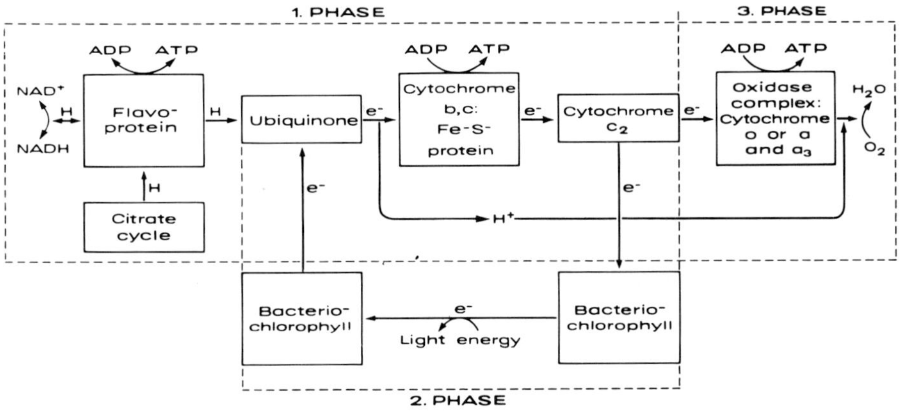

Fig. 36. Non-sulfur purple bacteria like **Rhodopseudomonas** are capable of both aerobic respiration and aerobic photosynthesis. The part of the electron transport chain which they use for respiration differs very little from that in **Paracoccus** or mitochondria. However, there is an additional "side track": Cytochrome $\underline{c}_2$ can transfer its electrons either to cytochrome oxidase or to bacteriochlorophyll: the e^- donated to the chlorophyll can be activated by sunlight, thereafter they flow via ubichinone and cytochromes $\underline{b}$ and $\underline{c}_2$ back to the bacteriochlorophyll. It is suggested here that first the short, anaerobic respiratory chain developed (phase 1), then anaerobic and later aerobic photosynthesis (phase 2), and that then the aerobic, long respiratory chain (phase 3) evolved. Finally, some descendants of the photosynthesizers (e.g. **Pseudomonas, Paracoccus, Beggiatoa, Leucothrix**) lost the photosynthetic apparatus and became the recent aerobically respiring prokaryotes (adapted from Dickerson 1980).

membranes the redox systems of the respiratory chain. The divergence of three major lines - the ur-karyotes, the archaebacteria, and the eubacteria - presumably took place with the beginning of development of the first cytochromes of the b type. The further evolution of archaebacteria to the eubacteria would then have occurred through further development of redox systems, e.g. cytochrome c, rhodopsin and cytochrome a, in the cell membrane. Before the incorporation of the last redox system, cytochrome a_3 (cytochrome oxidase), bacterial chlorophyll was integrated with a reverse cyclic electron flow (cf. Dickerson 1980; cf. also Fig. 36); this corresponds to its redox potential. The recent respirers are believed to have lost their chlorophylls later (Broda 1975).

On the basis of homologies among eubacterial, archaebacterial and eukaryotic RNA polymerases, **Zillig et al.** (1985) concluded that the relationship was close between the acidophiles and the eukaryotes on the one hand, and the methanogens and halophiles on the other. **Lake et al.** (1985), based on their three-dimensional analysis of ribosome fine structure, set the thermoacidophiles in their own group, the eocytes, removing them from the archaebacteria and setting them closer to the eukaryotes. They also separated the halophiles from the archaebacteria and included them and the eubacteria in the new group, the photocytes. However, because of the doubtfulness of morphological criteria as the sole phylogenetic indicators, these last divisions are disputed (Fewson 1986).

VI. References

Broda, E. (1975): *Evolution of bioenergetic processes*. Pergamon Press New York.

Dayhoff, M.O. (1972): *Atlas of protein sequence and structure*. National Biochemical Research Foundation, Washington D.C.

Dickerson, R.E. (1980): *Cytochrome c and the Evolution of Energy Metabolism*. Scientific American 3 98-110.

Doolittle, W.F., Bonen, L. (1981): *Molecular sequence data indicating an endosymbiotic origin for plastids*. In: Fredrick, J. (ed.): Origins and evolution of eukaryotic intracellular organelles. Ann. N.Y. Acad. Sci. 361 248-256.

Dose, K., Rauchfuß, H. (1975): *Chemische Evolution und der Ursprung lebender Systeme*. Wiss. Verlagsgesellschaft Stuttgart.

Fewson, C.A. (1986): *Archaebacteria*. Biochemical Education 14 103-115.

Kandler, O. and Zillig, W. (eds.; 1986): *Archaebacteria '85. Proceedings of the EMBO*, Gustav Fischer Verlag Stuttgart, New York.

Kaplan, W. (1978): *Der Ursprung des Lebens. Biogenetik, ein Forschungsgebiet heutiger Naturwissenschaft.* 2nd edition; Georg Thieme Verlag Stuttgart.

Keniry, M.A., Gutowsky, H.S. and Oldfield, E. (1984): *Surface dynamics of the integer membrane-protein bacteriorhodopsin.* Nature 307 383-386.

König, H. (1986): *Biologie der Archäbakterien.* Biuz 3 71-82.

Lake, J.A., Clark, M.W., Henderson, E., Fay, S.P., Oakes, M., Scheinman, A., Thorner, J.P. and Mah, R.A. (1985): *Eubacteria, halobacteria, and the origin of photosynthesis: the photocytes.* Proc. Natl. Acad. Sci. USA 82 3716-3720.

Rahmann, H. (1972): *Die Entstehung des Lebendigen.* Gustav Fischer Verlag Stuttgart, New York.

Remane, A. (1952): *Die Grundlagen des natürlichen Systems, der vergleichenden Anatomie und der Phylogenetik.* Akademische Verlagsgesellschaft Geest und Portig, Leipzig.

Schön, G. (1978): *Mikrobiologie.* Herder Verlag Freiburg.

Schwemmler, W. (1984): *Reconstruction of cell evolution: A periodic system.* CRC Press Boca Raton, Florida.

\- (1985): *Endocytobiosis formation: macromechanism of cell evolution.* Endocyt. C. Res. 2 191-211.

Stackebrandt, E. (1986): *Das hierarchische System der Eubakterien: Probleme und Lösungsansätze.* Forum Mikrobiol. 5 255-260.

Stetter, K.O. and Zillig, W. (1985): Thermoplasma *and the Thermophilic Sulfur-Dependent Archaebacteria.* In: Woese, C.R. and Wolfe, R.S., eds., Archaebacteria, Vol. 8, 85-170. In: The bacteria: a treatise on structure and function; Academic Press New York.

Woese, C.R. (1981): *Archaebacteria.* Scientific American 6 98-122.

Woese, C.R. and Fox, G.E. (1977): *The concept of cellular evolution.* J. Mol. Evol 10 1-6.

Zillig, W., Schnabel, R. and Stetter, K.O. (1985): *Archaebacteria and the origin of the eukaryotic cytoplasm.* Curr. Top. Microbiol. Immunol. 114 1-18.

Third Subchapter Eucyte

I. Phenomenon

The human body consists of many trillion cells (10^{13}-10^{14}). There are well over 200 types of cells in the body which carry out various functions. Some produce secretions, others form muscles, still others form hairs or bones. In their fundamental structure, however, they are all the same, and also correspond to the cells of protista, animals, fungi and plants. All animal and plant cells contain a membrane-bounded nucleus and for this reason, these organisms are called **eukaryotic cells** or "**eucytes**". The eucytes usually form into multicellular systems which arise through precisely regulated cell cleavage (mitosis, meiosis). In addition, eucytes are highly compartmentalized into separate reaction areas. The **compartments** have different functions (Fig. 37), such as heredity (nucleus), fermentation (cytoplasm), respiration (mitochondrion), photosynthesis (chloroplast), and a motility complex (i.e. flagellum[1]), the basis of which, however, is membrane-free, and therefore not a true compartment. These spaces are provided by genetically more or less independant cell organelles or organelle complexes. Only the nucleus-cytoplasm is (conditionally) genetically autonomous. Mitochondria and chloroplasts are semiautonomous organelles, whereas the only nucleic acids probably belonging to the motility apparatus are transfer RNAs (Hartmann 1975).

The question now arises as to the origin of the DNA- or RNA-containing eucyte organelles. Similarities in the genetic apparatus and metabolism of prokaryotes and eukaryotes indicate that the more highly developed eucyte must have evolved from the more primitive procyte. Two alternative hypotheses concerning the nature of this evolutionary process are currently discussed.

---

[1] of tubulin protein (cilium pattern (9+2) = eukaryotic flagellum = mastigion = euflagellum)

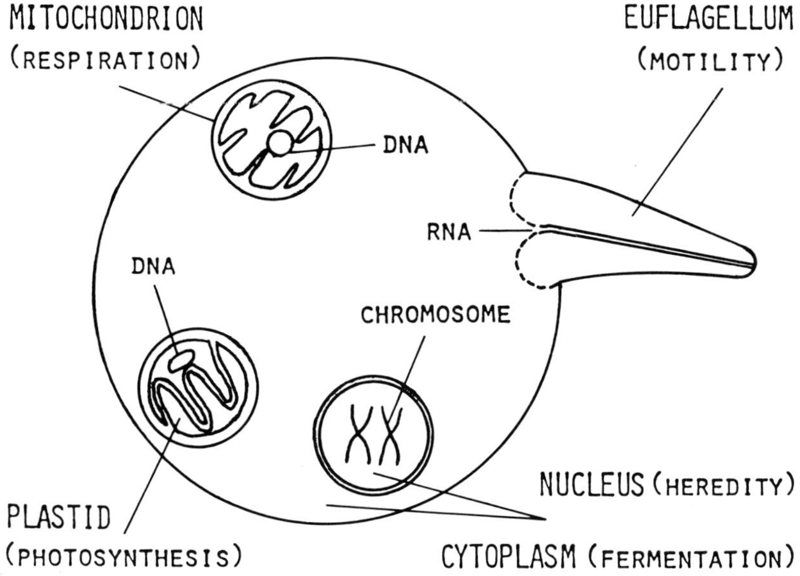

Fig. 37. Combined scheme of the connected, more or less independent reaction spaces in the eucyte (Schwemmler 1985).

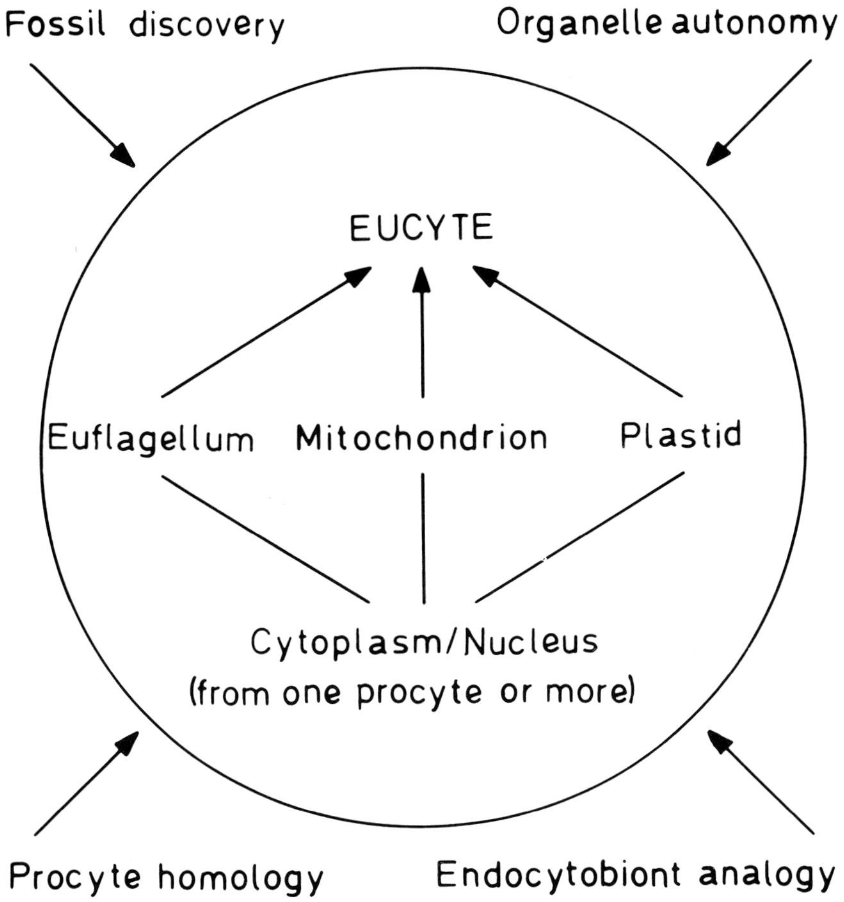

Fig. 38. Alternative hypotheses, methods of analysis, and fossils concerning eucyte evolution.

II. Hypotheses

The **compartmental** or **classical cell hypothesis** postulates that the eucyte is an elemental organism which evolved from a single procyte ancestor (Mahler and Raff 1972). The metabolic processes of the ancestral procyte, such as fermentation, photergism, respiration and photosynthesis, became compartmentalized through an endogenous process which incorporated the corresponding duplicated genes into the organelles.

According to the **endocytobiotic cell hypothesis** (an extension of the series endosymbiosis hypothesis: Taylor 1974), the eucyte developed from endosymbiosis of at least four different procytes (e.g. Schimper 1883, Margulis 1970, Schwemmler 1984). The procytes were representatives of the four metabolic types: fermenters, photergers (for definition, see p. 71), respirers and photosynthesizers. The essential difference between the two hypotheses is the origin of the extranuclear eucyte DNA: either from the endogenous (nuclear) or from exogenous (endocytobiotic) DNA[1].

III. Data

To verify or falsify one of the alternative hypotheses, various methods have been used: the analysis of microfossils and of genetic autonomy of the DNA- or RNA-containing organelles, and the study of their analogies with recent endocytobionts, and of homologies with procytic systems. In the following, the experimental data from these various studies will be presented briefly, to determine the extent to which they support one or the other of the alternative hypotheses. It can be assumed that whatever supports the one will in principle refute the other (Fig. 38).

---

[1] Arguments for the recently postulated symbiotic origin of DNA-free cell-organelles such as peroxisomes and hydrogenosomes, see Cavalier-Smith 1987b.

A. Cell fossil discovery

The interpretation of eucyte fossils is problematic. There is a lack of adequately preserved substructures and cell-organelles, which usually makes unequivocal identification of the fossils difficult; in some cases it is not even possible to clearly distinguish eucyte and procyte fossils (Schopf and Zeller-Oehler 1976).

The supposedly oldest eucyte fossils were found in the $2x10^9$ - $1.9x10^9$ years old Gunflint formation in North America. These are spherical single cells, 20 μm in diameter. They are relatively well preserved through silicification, and can be studied in three dimensions by taking thin slices. Some paleontologists regard these fossils as the first evidence of **euflagellated eucytes**, though this is dubious; they could easily all be cyanobacteria (for a review see e.g. Schwemmler 1984). The Crystal Spring layer of North America yielded $1.5x10^9$- $1.2x10^9$ year old multicellular, filamentous microfossils which are supposed to be primitive fungal hyphae (cited in Kaplan 1978). According to these very doubtful interpretations, **mitochondria-containing eucytes** would have existed by this time at the latest. So far, the oldest fossils which probably contain **plastids** have been found in the $1.4x10^9$ - $1.2x10^9$ years old Beck Spring layer in California, or in the Nonesuch formation in North America and in the Bitter Spring layer in Australia, both about 10^9 years old. These oldest fossils are spherical cells or multicellular filaments, 14 - 18 μm in diameter and 15 - 100 μm long. These cell structures are similar to those of green algae or cyanobacteria.

If one assumes that the above listed interpretations of the fossils are correct (which is very uncertain), the organelles will appear in the following order in the course of evolution:

Euflagellum --> Mitochondrion --> Plastid.

It is now interesting to observe the degree of genetic autonomy of these three organelles in relation to this hypothetical phylogenetic order.

B. Organelle autonomy

Plastid. Like the other types of plastids (proplastid, amyloplast, leukoplast, etioplast, chromoplast), chloroplasts contain their own genetic system, a ring-shaped DNA molecule with contour lengths of about 50 μm. This corresponds to a molecular weight (MW) of about 10^8 daltons, which is much less than that of the DNA in certain bacteria like mycoplasmas and rickettsias. With 250,000 to

300,000 base pairs, the chloroplast genome can encode the synthesis of about 200 different proteins with an average MW of 40,000. However, as activity measurements have shown, only about 20% of the genetic information is active: this means the synthesis of more than 50 macromolecules like proteins (5-10% of the proteins of the inner membrane, a few enzymes of photosystem I, those of the electron transport chain, and subunits of ribulose 1,5-biphosphate carboxylase) and nucleic acids[1] (3 ribosomal RNAs, over 20 t RNAs). All the remaining organelle components are encoded in the nucleus and nearly all of them are synthesized in the cytoplasm. If one compares the amount of plastid-endogenous proteins and nucleic acids to the minimum requirements of a procyte for autonomous morphogenesis, estimated by Kaplan (1978) to be about 100 macromolecules, then a plastid encodes less than 50% of its own essential molecules. If one assumes as a further great simplification that the percentage of plastid-endogenous macromolecules is equal to the degree of genetic autonomy, then the plastids are less than 50% autonomous and more than 50% dependent on the nucleus. Thus plastids have an endogenous genetic apparatus, a protein-biosynthesis system and a metabolism, but essential parts of these are encoded in the eucyte nucleus and synthesized in the cytoplasm. Genetically speaking, the plastids are thus **semiautonomous organelles**, which, however, multiply by division.

Mitochondrion. Fine-structure and biochemical studies have shown that mitochondria of animals and fungi also contain a circular DNA molecule between 5 and 25 μm contour[2]; however, its MW of 10^7 daltons is very much less than that of primitive procytes. The mitochondrial genome is fully active, and with its 15,000 to 30,000 base pairs is capable of synthesis of about 30 different macromolecules, such as proteins with an average MW of 20,000 (5-10% of the

---

[1] So far two chloroplastid DNAs have been fully sequenced.

[2] much more variable in fungi than in animals.

Fig. 39. Characterization of the genetic system and of the postulated phylogenetic age of euflagella, mitochondria and plastids.

BP, base-pairs.

* Mitochondrial DNA of plants is much larger.

	EUFLAGELLUM	MITOCHONDRION	PLASTID
DNA	---	~10^7 d ~30,000 BP* ~30 proteins/RNAs	~10^8 d ~300,000 BP ~50 proteins/RNAs
RNA	endogenous ?	endogenous	endogenous
Postulated GENETIC AUTONOMY	(0% ?)	<30%	<50%
Postulated NUCLEAR DEPENDENCY	(100% ?) <	>70 % <	>50%
Postulated AGE (years)	$2x10^9$ >	$1.5x10^9$ >	$1x10^9$

proteins of the inner membrane: ATPase, subunits of cytochromes, insoluble respiratory chain proteins, a few ribosomal proteins) and nucleic acids (three ribosomal RNAs, over twenty tRNAs). The great majority of the organelle components is encoded by the nucleus and most components are synthesized in the cytoplasm. Thus the mitochondrion has much less than 30% genetic autonomy, i.e. it depends more than 70% on the nucleus. Like the plastids, mitochondria are **semiautonomous organelles**, which can only arise through division of other mitochondria.

Euflagellum. Most of the membrane-enclosed euflagella and cilia of eucytes consist of two central microtubules and nine pairs of peripheral microtubules, that is, the basic pattern is (9+2). They are anchored in the cytoplasm by and generated from basal bodies (kinetosomes), which are not enclosed in a membrane and which have the microtubule pattern (9+0). The membrane-free centrioles, which play a role in mitosis and meiosis, also have the pattern (9+0) and are homologous to the basal bodies; they can also arise from basal bodies by reduplication. Nevertheless, no discovery of endogenous DNA in centrioles and basal bodies has yet been substantiated. The isolation of endogenous euflagellar DNA with a MW of about 10^6 daltons has been reported. However, this is disputed with the argument that the DNA is certainly an artefact of incomplete separation from the mitochondria, which are always closely associated with the motility apparatus. Furthermore, the base sequence of the isolated DNA was found to be largely the same as that of mitochondrial DNA. In other words, the dependence of the euflagellum on the nucleus is 100%.

If one now compares plastids, mitochondria and euflagella with respect to the loss of their demonstrated or postulated genetic autonomy, one obtains a series of decreasing autonomy, which is in exactly the reverse order of their postulated phylogenetic age (Fig. 39):

Plastid --> Mitochondrion --> Euflagellum

This context supports the endocytobiotic cell hypothesis according to which the most recent, and thus the least integrated, reduced or modified organelles (like the plastids) would also have the greatest degree of genetic autonomy. The presumably phylogenetically oldest organelles (euflagella) would be most degenerate with regard to their genetic autonomy, i.e. most strongly integrated into the cell. However, the high genetic dependence of organelles on the nucleus is put forward as an argument for the classical cell hypothesis, and against the endocytobiont hypothesis. It is a weak counterargument, because a phylogenetic development from endocytobiont to cell-organelle requires increasing integration of the original endocytobiont into the cell. For example, the DNA polymerases

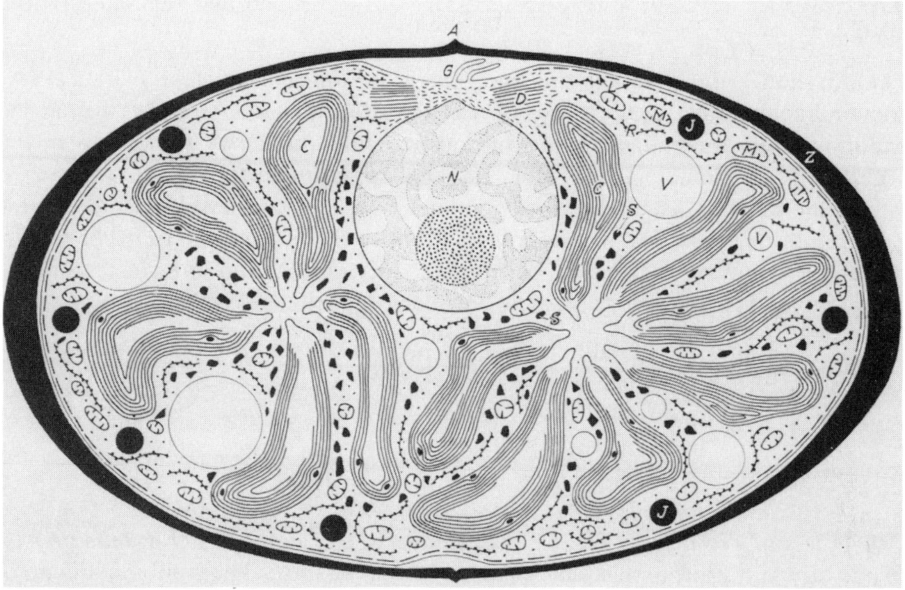

Fig. 40 a. Pattern for organization of endocytobiosis between **Glaucocystis nostochinearum** and a bluegreen bacterium or a cyanelle of unknown nature (Schnepf and Brown 1971).

A = equator	C = cyanelle*	G = euflagellum	J = osmophilic globuli
N = nucleus	M = mitochondrion	R = endoplasmic reticulum	
V = vacuole	Z = host cell wall	S = starch grains	

Fig. 40 b. Section through the membrane of a **cyanelle** (C): symbiont-derived residual cell wall (W) in the form of murein between host membrane (B) and symbiont plasmalemma (P); according to Schnepf and Brown 1971.

* Some authors regard these cyanelles as true chloroplasts.

responsible for replication of organelle DNA are encoded and regulated by the nucleus.

The question arises whether the low genetic information content of the DNA-containing organelles is a result of **transfer** of DNA units from the endocytobiotic organelle into the nucleus, so that the DNA remaining in the organelle is a residue left over after adoption of an endocytobiont. It is therefore interesting to examine whether the genomes of recent endocytobionts also show a degree of degeneration or integration into the host's genome. In other words, has there been gene transfer between endocytobiont and host cell genome comparable to that which we suspect between organclle and cell nucleus?

Fig. 41 a. Embryonic host cell or bacteriocyte of the leafhopper **Euscelis incisus** filled with type **a** protoplastoid primary endocytobionts*. The symbionts contain many optically dark inclusion bodies; the nucleus of the host cell is polyploid (Schwemmler 1984).

a_v = primary symbiont type* **a** in the vegetative form
$A_{1,2}$ = various types of inclusion bodies
HS = heterochromatin structure of the cell nucleus
M_i = mitochondrion
N = host cell nucleus
t_i = auxiliary symbiont* type **t** in the infection form

Fig. 41 b. Section through the membrane of the vegetative form (SV) of the primary symbiont type **a**, showing the membrane layer M_3 formed by the host cell cytoplasm (Cy), the symbiont membrane layer or symbiont plasmalemma M_1, and between M_3 and M_1 another layer, M_2, presumably symbiotic residual cell wall (Louis and Laporte 1969).

* According to **Müller** (1949), primary symbionts are endocytobionts vital for the host. They are the only ones in leafhoppers appearing without other kinds of endocytobionts, whereas auxiliary endocytobionts can only exist in association with primary endocytobionts. The endocytobiotic infectious forms represent those stages passing on the endocytobionts extracellularly from one host generation to the next; the vegetative forms represent stages reproducing intracellularly.

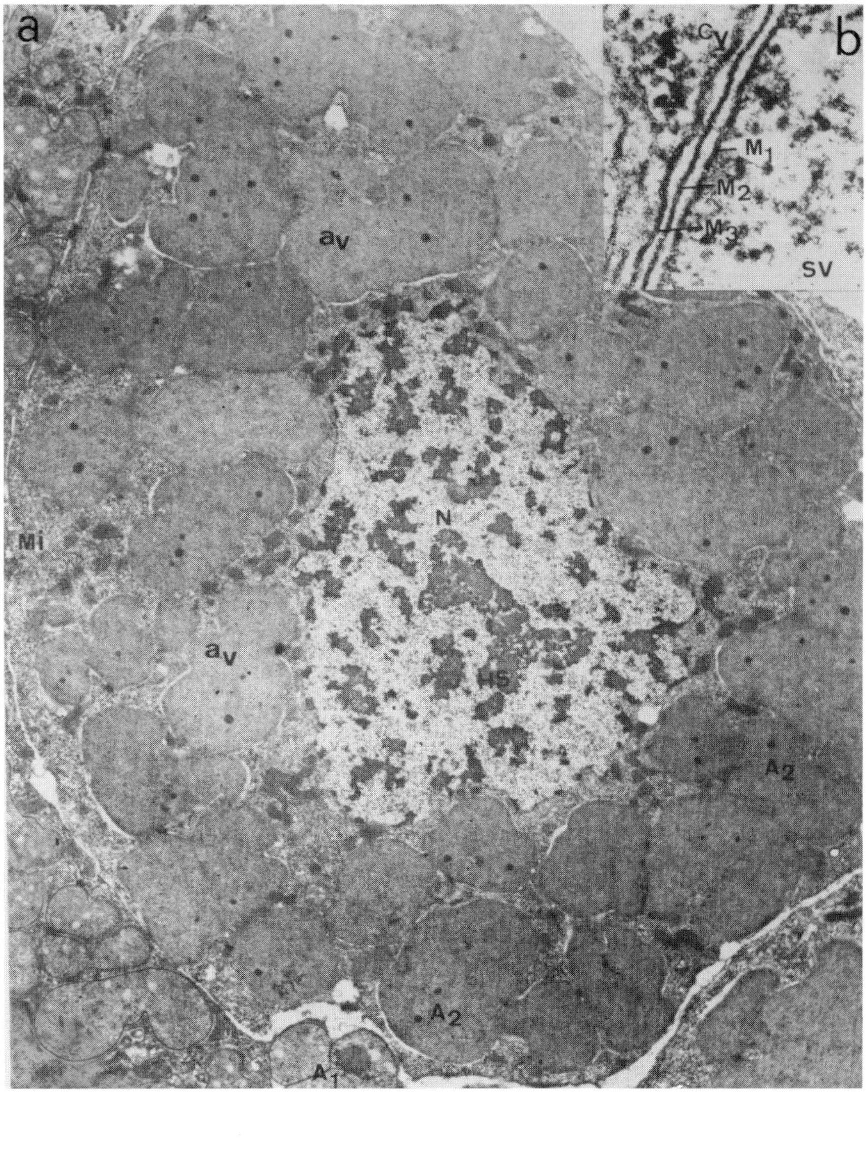
a
b
cv
M1
M2
M3
SV
av
N
Mi
av
HS
A2
A2
A1

C. Endocytobiont analogy

Plastid. Syncyanoses are symbioses between bluegreen bacteria (bluegreen "algae") and zooflagellates or fungi or plastid-free euflagellates. A spectrum exists, from loose associations, which can be dissolved at any time, to vital endocytobioses, in which the symbiont carries out certain functions vital to the host cell (literature review, e.g. Schenk 1977). At the beginning of the chain there are the phylogenetically young endocytobioses between the lower fungus **Geosiphon pyriforme** and bluegreen bacteria of the genus **Nostoc**. The fungus constantly takes up bacteria from the soil and releases them again. It surrounds the bacteria with its own cell membrane in phagocytosis vacuoles. The bluegreens within the host eucyte are not visibly changed; there is no degeneration of cell wall. The symbionts can be isolated from the host cell at any time and may be cultured independently. The host can also exist without the bluegreens. Thus this endocytobiosis[1] is not essential for either partner; it is facultative.

Close to the other end of the phylogenetic sequence is the endocytobiosis between the plastid-free host **Glaucocystis nostochinearum**, and bluegreen bacteria of unknown nature (cf. the host **Cyanophora paradoxa**). In this case the bluegreens, too, are enclosed in "phagocytosis vacuoles" (Fig. 40, a). In **Glaucocystis**, the membrane around the phagocytosis vacuole contains cholesterol, and is comparable in composition to the outer plastid membrane. The plasmalemma of the symbiont, in contrast, has the usual procyte complement of enzymes, and thus is similar to the inner plastid membrane. In both these bluegreens and in plastids, pigmented thylakoids are formed by invagination of the inner membrane; these are the sites of photosynthesis. The cell wall of the bluegreens in this symbiosis has almost disappeared, leaving only traces behind (Fig. 40, b). In the space between the phagocytosis vacuole membrane and the symbiont's plasmalemma, residual cell wall can be detected in the form of murein. This space thus corresponds to that between the inner and

---

[1] Explanation of endocytobiosis, see p. 203

Fig. 42. Analogies of mitochondria and protoplastoid endocytobionts in leafhoppers (Schwemmler 1984).

* Chelates are complex compounds of carbohydrates with inorganic ions.

P R O T O P L A S T O I D (endocytobiotic bacterium)	M I T O C H O N D R I O N (eucyte organelle)
- Outer symbiont membrane of cytoplasmic origin	- Outer mitochondrial membrane possibly of cytoplasmic origin (e.g. enzyme components)
- Symbiont plasmalemma	- Inner mitochondrial membrane (often with cardiolipin)
- Cell wall residues in the space between outer symbiont membrane and symbiont plasmalemma	- No cell wall residues in the space between outer and inner mitochondrial membrane
- Invagination of the inner symbiont membrane (e.g. membrane bodies, mesosomes, division membrane)	- Invagination of the inner mitochondrial membrane (e.g. cristae/tubuli, division membrane)
- Symbiont matrix (circular chromosomes ? : 10^8 daltons at most, 70S ribosomes etc.)	- Mitochondrial matrix (circular chromosome: ca. 10^7 daltons, 70S ribosomes etc.)
- Inclusion body: chelates* (Ca^{2+}, PO_4^{3-}, proteins)	- Granula: hydroxyapatite (Ca^{2+}, PO_4^{3-}, proteins)
- Regulation of pH, cell rhythm and osmotic pressure for the host cell	- Regulation of pH, cell rhythm and osmotic pressure for the eucyte
- Infectious vegetative forms	- Compact (condensation), hyaline (orthodox) forms
- Reproduction dependent on host cell (semiautoreproductive)	- Reproduction dependent on eucyte (semiautoreproductive)

outer plastid membranes. In this syncyanosis, the bluegreens are so firmly integrated and so highly degenerated, that they are no longer capable of existing outside of the host cell. The host is also unable to live without the symbionts which provide it with photosynthetic CO_2 fixation. The symbiosis is thus obligatory for both partners, and they have attained a high degree of mutual dependence. In the unicellular, eukaryotic alga **Glaucosphaera vacuolata**, cell cleavage is even synchronized with that of the symbiotic bluegreens (compare similar interrelations in chloroplasts). The high degree of integration and degeneracy suggests a gene transfer between the two systems. Therefore it seems justifiable to place such endocytobiotic bluegreen bacteria on the same level as chloroplasts. In order to emphasize their organelle character, such endocytobiotic bluegreens are also called **cyanelles**.

Mitochondrion. Leafhoppers as a group (Hemiptera, Homoptera, Cicadina) display a phylogenetic series of increasingly integrated and degenerate endocytobionts (Schwemmler 1984). The phylogenetically recent, rickettsia- or eubacteria-like **accessory symbionts** are usually taken up by unspecific insect cells. They are largely independent of their host eucytes. There is no existential reduction, especially of the cell wall, by the host. Host cell and symbiont are separately viable. Their symbiosis is thus facultative. The phylogenetically old protoplastoid[1] **auxiliary** and **primary** endocytobionts of the leafhoppers, by contrast, are always maintained in special host cells (Fig. 41, a). When they are taken in by these cells, the endocytobionts are surrounded by a host membrane. Only traces of cell wall residues can be found in the symbionts (Fig. 41, b; Louis and Laporte 1969). They have also come under the influence of the host cell with respect to energy metabolism. Outside the host cell, they are no longer able to regulate their own reproduction and soon die. Thus these endocytobionts have partially lost their autonomy. However, this symbiosis is also obligatory for the host; if it is interrupted, the insect dies within a few days. The endocytobionts are therefore passed on from one generation of leafhoppers to the next, via the egg. If the transfer of symbionts together with the egg is interrupted spontaneously or artificially, one obtains non-viable partial embryos. These consist of head and thorax, but have no abdomen. The endocytobionts also synthesize metabolites for the host from its metabolic waste products. In this way they regulate the pH and osmotic pressure in connection with the formation of inclusion bodies. They probably also influence certain cell rhythms in the host's energy metabolism. Thus these endocytobionts have functions which are qualitatively and quantitatively entirely comparable to those of mitochondria. The DNA molecular weight of the endocytobionts is about $2\text{-}3 \times 10^7$ daltons, which is only slightly

---

[1] Protoplastoids are a newly recognized group of intracellular procytes quite similar in many respects to the mycoplasmas and rickettsias (Schwemmler 1984).

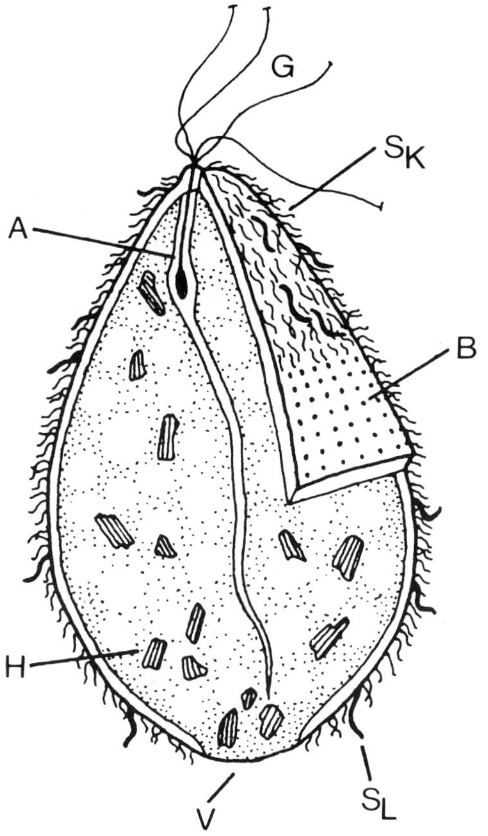

Fig. 43 a. Organization of the symbiosis of the host flagellate **Mixotricha paradoxa** with episymbiotic bacteria and spirochetes.

A	=	axostyle	G	=	euflagellum	S_L	=	long spirochete
B	=	bacterium	H	=	wood	S_K	=	short spirochete
Mu	=	filament	P	=	plasmic extension			
V	=	digestion zone						

higher than that of mitochondria, and far less than that of primitive procytes. This suggests that gene transfer between endocytobionts and host cells has occurred. Fig. 42 gives a list of the most important analogies between protoplastoid leafhopper endocytobionts and mitochondria (compare p. 184).

In the giant amoeba, **Pelomyxa palustris**, there are endocytobiotic bacteria instead of mitochondria (cf. also the primarily or secondarily mitochondria-free flagellate **Mixotricha**). Like mitochondria, these bacteria apparently occur in two types of structure, e.g. in a small, optically dense ("compact form") or in a larger, optically transparent form ("hyaline form"). The larger forms are enclosed within a tubule network comparable to the endoplasmic reticulum. The bacteria apparently carry out respiration, and thus take over the function of the missing mitochondria (Whatley 1976). In addition, endocytobiosis appears to be obligatory for both partners. The symbionts have partially lost their autonomy to the host eucyte. However, the host cell is also dependent on the endocytobiont. This interdependence possibly originates in gene transfer between the two systems. The integration has reached a level comparable to that of the mitochondrion.

Euflagellum. Many protista form more or less intense symbioses with mobile bacteria (Margulis 1970). For example the zooflagellate **Pyrsonympha** harbours episymbiotic spirochetes. These are fixed on one end to the surface of the euflagellate. This gives the impression of a set of cilia, or of additional euflagella. The first tendency toward dissolution of the cell wall can be seen at the contact zone. These protists were therefore mistakenly placed at first in the group of the **Hypermastigina**. In the zooflagellate **Mixotricha paradoxa**, which lives in the intestine of termites, the episymbiosis with spirochetes has proceeded still further. Electron microscopic studies show that symbiotic bacteria of unknown nature are associated in a regular pattern on the surface of the euflagellate (Fig. 43, a). Symbiotic spirochetes "insert" into the bacteria through a cytoplasmic extension of the host's surface, as in a type of euflagellar basal body (Fig. 43, b).

Fig. 43 b. Section through the surface of the host euflagellate (E) Mixotricha paradoxa; where episymbiotic bacteria (B) are associated in a regular pattern, the spirochetes (S_K) "insert" into the bacteria. The bacterium-spirochete complex is fixed to a cytoplasmic extension (P) from the host surface.

Fig. 43 c. Cross section of an episymbiotic spirochete showing the typical spirochete filament pattern (Fi) in the protoplasmic cylinder (a,b and c according to Grimstone 1964).

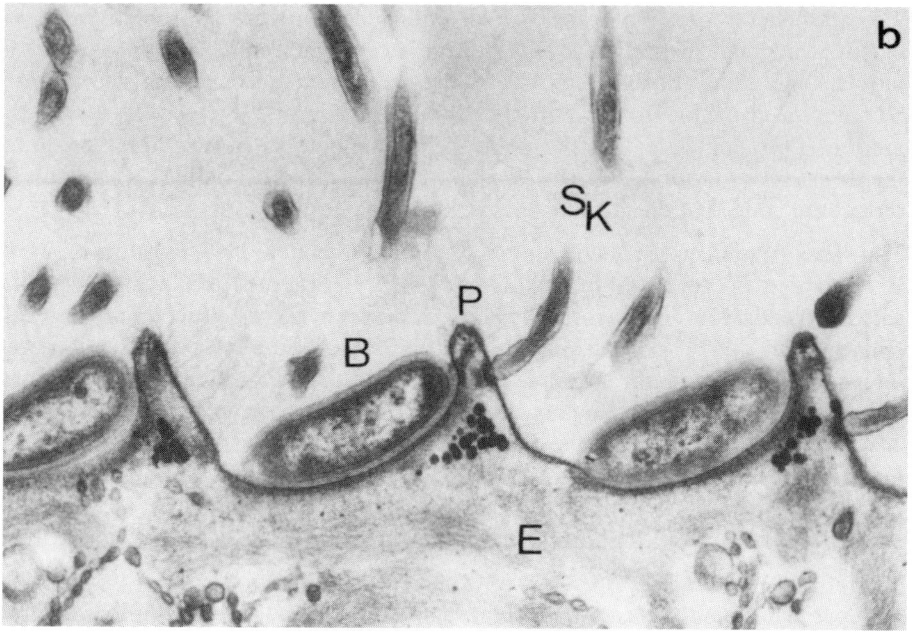
b
SK
P
B
E

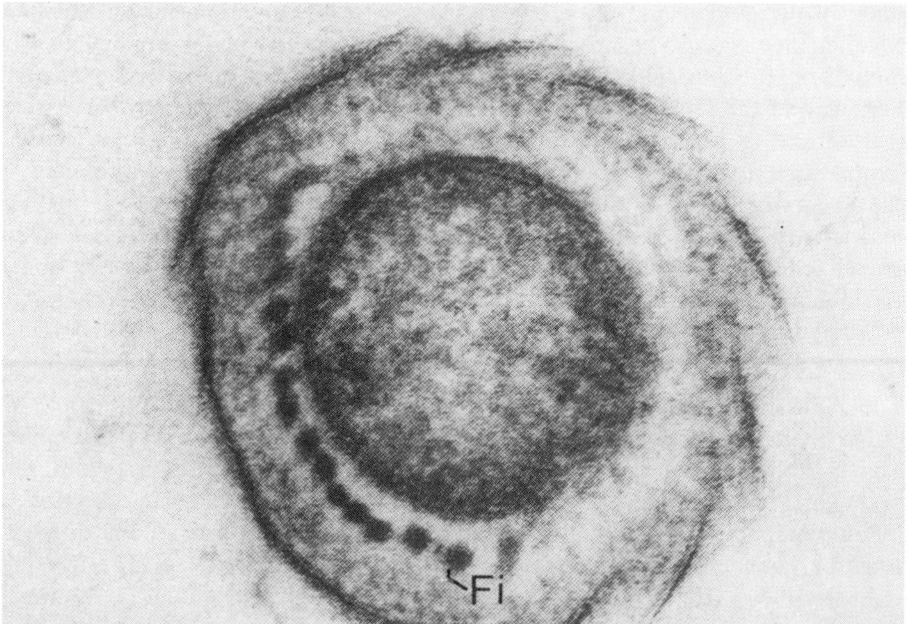
Fi
c

The spirochetes are about the size of euflagella, and move using periplasmic bacterial flagella. In this case, a flagellum is composed of a bundle of about 17 motile longitudinal fibrils (Fig. 43, c). The bacterium-spirochete-host-cytoplasm complex is also able to beat in coordinated fashion. It exclusively serves the motion of the flagellate; the four polar euflagella merely act as rudders to control the direction of motion. For this reason, the episymbiotic complex was formerly thought to consist of eucyte cilia.

The latter symbiosis is not essential for either the host cell or the symbionts. Each partner is capable of independent existence (Margulis 1981). Even with the small projection of host cytoplasm, the symbionts are not intracellular, but only episymbiotic. They also display no recognizable signs of degeneration and integration. Their membranes are conserved, and their genetic information is not reduced. The euflagellum, by contrast, is not entirely surrounded by membrane, at least not in the region of the basal body, and DNA is either not present, or it has been largely or entirely integrated into the nuclear genome. Therefore it does not (yet) seem possible, given the present state of knowledge, to compare the eucyte euflagellum with recent endocytobionts, due to a lack of suitable final stages.

Comparisons with recent endocytobionts thus reveal levels of integration comparable to those of the plastids and mitochondria, but not comparable to those of the euflagellum. In each case, it is actually only a matter of definition whether they are still regarded as endocytobionts or already as organelles. The mutual interactions with the nucleus are genetically controlled and regulated, here as with the DNA-containing organelles. This goes so far in another system, that of the host ciliate **Paramecium aurelia** with its endocytobiotic bacteria of the "kappa" type, that the nuclear genes of the host cell influence the reproduction of the endocytobiont (Görtz 1980). The ciliate **Euplotes aediculatus** is actually incapable of cell division if it loses its endocytobiotic bacteria. All of these phenomena may require an exchange or transfer of genes between host and symbiont. The genetic correlations between host eucytes and endocytobiotic procytes are still not well understood. Further and intensive studies are therefore not only desirable, but necessary for a deeper understanding of the total complex.

The above observations principally all support the endocytobiotic cell hypothesis rather than the classical cell hypothesis, although it is necessary to distinguish exactly between the individual organelles. In spite of intensive investigation, no examples are known which would support a stepwise intracellular evolution of DNA-containing cell organelles. Quite the contrary, any sort of intermediate stage between procytes and eucytes is completely missing. At best, one could refer to the nuclei of **dinoflagellates**, which arose endogenously, however, according to both hypotheses. But the lack of transitions between procytes and eucytes favours the endocytobiotic hypothesis over the classic cell hypothesis. If

	BACTERIAL ENDOCYTOBIONTS	
	Facultative	Obligate
Analogous to EUFLAGELLUM	Spirochete episymbiosis Motility organelle Host: euflagellate	?
MITOCHONDRION	Rickettsia companion endocytobiosis Nutritional complement Host: aphids, leafhopper	Protoplastoid auxiliary / primary endocytobiosis Glycolysis / respiratory organelle Host: aphids, leafhopper
PLASTID	Bluegreen endocytobiosis (syncyanosis) Nutritional complement (assimilates) Host: unicellular fungus	Bluegreen endocytobiosis (syncyanosis) Photosynthesis organelle (cyanelle) Host: plastid-free alga

Fig. 44. Analogies between endocytobionts and DNA-containing cell organelles.

the eucyte evolved through formation of endocytobiosis, "phylogenetic leaps" without intermediate stages would be expected instead of a gradual, continuous evolution caused by mutation and selection.

As the recent endocytobioses show, the phylogenetic derivation of the DNA-containing organelles from prokaryotic endocytobionts is possible in principle (Fig. 44). However, the question arises as to whether the organelles are actually homologous to recent procytes. Data from comparative sequence analyses will be useful here. However, even base-pairing or hybridization experiments can serve as criteria of homology. Under certain conditions, double-stranded DNA can be converted into single-stranded molecules. In this form, it may pair with other single-stranded DNA or RNA. RNA can also be directly paired with other RNA. The number of base pairs formed in the hybrid molecules can be estimated from the "melting temperature" (the temperature at which the double strands separate); the "melting temperature" is thus an expression of the degree of homology or relatedness of the two hybridized nucleic acids.

D. Procyte homology

Plastid. As hybridization experiments show, plastid DNA from **Euglena** is only about 1% homologous to cytoplasmic RNA from the same cell, up to 4.5% homologous to bacterial RNA, and up to 47% homologous to bluegreen bacterial RNA. This indicates that there is little relation between the chloroplast and the cytoplasmic system (Fig. 45). On the other hand, there are general similarities to procyte systems (e.g. antibiotic sensitivity, and genetic recombination through plastid fusion, as in **Chlamydomonas and Polytoma**). The relationship between chloroplasts and bluegreen bacteria is seen here to be especially close. In

---

Fig. 45. Homologies between DNA-containing cell organelles as well as endocytobionts and procytes, and analogies with the eucyte system (Schwemmler 1984).

* Exception: eucyte dinoflagellates with DNA molecules nearly histone-free, circular, membrane-associated.

** Not in archaebacteria; in microsporida: rRNA: 23S, 16S, and 70S ribosomes.

PROCYTE/CHLOROPLAST/MITOCHONDRION	EUCYTE (CYTOPLASM)
Genetic apparatus	
Circular chromosome: $<10^{12}$ daltons No histones (or low content) Associated with membrane Specific base composition: G-C $>$ A-T One starting point for replication Non-Mendelian genetics	Linear chromosome: $>10^{12}$ daltons Associated with histones. Not bound to inner nuclear membrane Specific base composition: G-C $<$ A-T Several starting points for replication Mendelian inheritance
Protein-biosynthesis	
Specific mRNA rRNA (23S, 16S, 5S) tRNA 70S ribosomes (50S, 30S) 52-55 specific ribosomal proteins Specific aminoacyl tRNA synthetases Protein synthesis begins with $tRNA^{N\text{-}F\text{-}Met}$ ** Rifamycin** sensitivity (transcription) Chloramphenicol sensitivity (translation)	Specific mRNA rRNA (28S, 18S, 5S ?)* tRNA 80S* ribosomes (60S, 40S) 70-84 specific ribosomal proteins Specific aminoacyl tRNA synthetases Protein synthesis begins with $tRNA^{Met}$ α-amanitin sensitivity (transcription) Cycloheximide sensitivity (translation)
Membrane components	
Special enzyme pattern Specific glycolipids and phospholipids (e.g. cardiolipin: not in plastids)	Special enzyme pattern Specific glycolipids and phospholipids (e.g. cholesterol)

addition, it should be mentioned that the base sequence of the individual ribosomal RNAs from chloroplasts is almost homologous to sequences in bluegreen bacteria (Fig. 46). They differ no more from one another than do those of some species of bacteria. In addition, components of the transcription and translation apparatus from chloroplasts and bluegreens can be exchanged in cell-free systems; examples are mRNA, tRNA and the associated amino-acid loaded enzymes (aminoacyl-tRNA synthetases). The physical and chemical properties of the four protein subunits of the chlorophyll-binding complex (of photosystem I) are the same as those from the bluegreens. Both subunits of the chromoprotein phycoerythrin from chloroplasts can even be exchanged for the corresponding subunits from bluegreens without loss of activity. These and other criteria (Margulis 1981) indicate that chloroplasts and bluegreen bacteria may derive from common procytic ancestors.

There are differences between the DNA-containing organelles and the cytoplasmic system of the eucyte, with regard to the transcription and translation apparatus and to metabolism; these systems are analogous. On the other hand, the organelles display close relations, or homologies, to recent procytes. This in turn agrees with the endocytobiotic cell hypothesis. Both hypotheses assume, however, that at the beginning of eucyte evolution, procyte conditions, e.g. 70S ribosomes, must have existed. Although the organelles were conservative, the eukaryotic system with its 80S ribosomes is thought to have arisen due to a different rate and direction of evolution.

Mitochondrion. The ribosomes of the mitochondrion, like those of bacteria, have a sedimentation constant of 70S, while the cytoplasm of most eucytes (exception: 70S for microsporida) contains only 80S ribosomes. Mitochondrial and bacterial 70S ribosomes, unlike the cytoplasmic ribosomes, consist for the most part of 50S and 30S subunits. The 50S subunit in bacterial and plant mitochondria (but not in animal or fungus mitochondria) has a 5S RNA, and the 30S subunit has a 16S RNA. The sequences of these ribosomal RNAs are homologous in plant mitochondria and bacteria (Fig 46). On the other hand, there is less homology to the corresponding cytoplasmic RNA (Fig. 44). Mitochondrial ribosomes can be functionally replaced by bacterial ribosomes in protein-biosynthesis, but not by cytoplasmic ribosomes. Both mitochondrial and bacterial protein-biosynthesis is initiated by the same N-formylmethionine transfer RNA, while eucyte protein-biosynthesis is initiated by a different tRNA, Met-tRNA. The messenger RNA required for protein-biosynthesis is synthesized on DNA by the enzyme RNA-polymerase. The function of the mitochondrial RNA-polymerase from the fungus **Neurospora** can be carried out in vitro by the RNA-polymerase from the bacterium **Escherichia coli**, but not by the specific eukaryotic RNA polymerase. Furthermore, protein-biosynthesis of mitochondria and bacteria can be blocked by the same antibiotics, e.g. chloramphenicol, streptomycin, erythromycin and tetracycline, which do not, however, affect the synthesis process on cytoplasmic

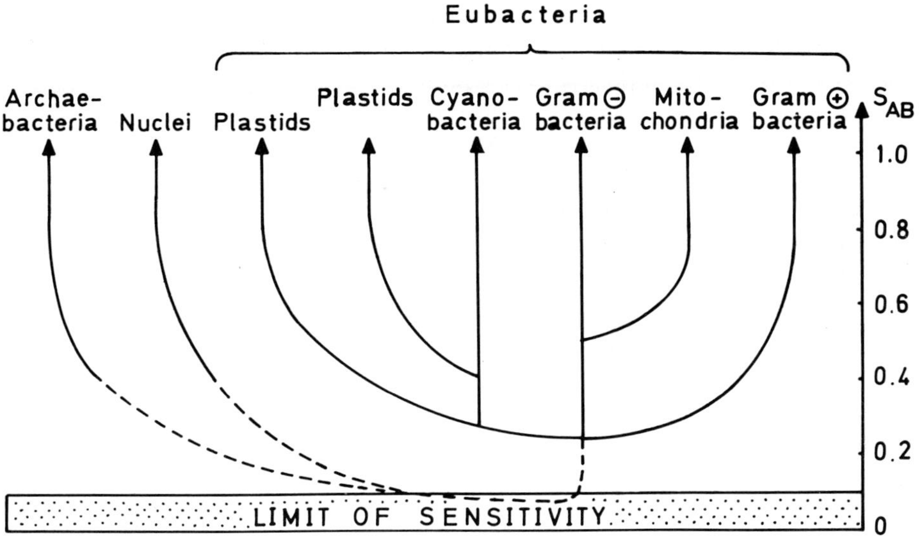

Fig. 46. Schematic phylogeny, based on 16S and 18S rRNA catalogues of cyanobacteria, plastids, bacteria (including archaebacteria), mitochondria and corresponding components of the nucleus-cytoplasm system of eukaryotic cells (Data from Woese and Fox 1977, Woese 1981, adapted from Doolittle and Bonen 1981). The similarity coefficient S_{AB}, a measure of the fraction of common sequences, is used here to indicate degrees of relationship among the organisms.

80S ribosomes. Conversely, antibiotics like cycloheximide inhibit cytoplasmic protein-biosynthesis, while that of mitochondria and bacteria is insensitive.

In the cases discussed above, the agreement is greater between mitochondria and bacteria than between organelles and cytoplasm of the same cell. Mitochondria show particularly strong relationships with certain types of bacteria; the amino acid sequence of the mitochondrial respiratory enzyme cytochrome c is clearly related to that of the respiring, non-sulfur purple bacteria. In addition, the mitochondria of the euflagellate **Polytoma**, for example, fuse with subsequent genetic recombination, as the corresponding bacteria are known to do. These criteria and others (Gray and Doolittle 1982) indicate the procyte nature of the mitochondrion. They make the derivation of these organelles from non-sulfur purple bacteria appear nearly certain.

Euflagellum. So far, DNA has not been isolated from euflagella, basal bodies or centrioles. Many results even exclude the existence of such DNA (Pickett-Heaps 1974). However, an euflagellar RNA is said to have been found, though not quantitatively prepared (Hartmann 1975). Thus comparative sequence analyses are presently not possible. **Margulis** (1975) postulates homologies between certain proteins of the eucyte motility apparatus and proteins of spirochetes. However, it is also postulated that the euflagellar apparatus could be derived from photergic bacteria of the halobacterial type with an additional (9+2) tubule pattern (Schwemmler 1984,1985: cf. Fig. 47 a). Until there is experimental confirmation, such claims remain working hypotheses, according to which the eucyte euflagellum and its (9+2) or (9+0) homologues (cilium, basal body, centriole, centromere?, microtubules) are derived from ancient **spirochetes** or photergic, spirochete-like bacteria with a (9+2) tubuli pattern.

There are other results in which there is no correspondence between DNA-containing organelles and the corresponding procytes, and these results are sometimes used to support the classic cell hypothesis. Differences in the DNA, the ribosomes and in the metabolism of bacteria and mitochondria on the one hand, and of bluegreen bacteria and chloroplasts on the other, are pointed out. According to **Arnold** (1977), such examples do not disprove the endocytobiotic cell hypothesis, since "plastids are certainly not bluegreen bacteria, and mitochondria are not bacteria. In the sence of the hypothesis, they only have a common origin. This goes back to a stage in the development of life which occurred before the existence of eukaryotic cells. Therefore one cannot expect comparative studies to reveal complete identity between prokaryotic organisms and eukaryotic organelles. It is rather a matter of detecting relics which indicate a common descent."

Other arguments against the endocytobiotic cell hypothesis refer to the differences in the DNA and RNA of mitochondria or plastids. For example, the mitochondria of plants have ribosomes of about 78S and circular DNA molecules 12-30 μm long. The mitochondria of higher animals, have DNA only 5 μm long and extremely small ribosomes of about 55S. On the other hand, only the plastids of **red algae** have phycoerythrin and phycocyanin, as bluegreen bacteria do; the chloroplasts of **green algae** do not have these proteins. However, these facts are not so much an argument against the endocytobiotic cell hypothesis as rather an argument for a **polyphyletic origin** of **mitochondrial** and **plastid endocytobioses**. On the other hand, a monophyletic derivation cannot be convincingly excluded.

IV. Reconstruction

A. The problem

In the reconstruction of eucyte evolution, the question as to the origin and nature of the presumed euflagellar symbiont with its (9+2) and (9+0) homologues remains open. **Margulis** believes that she has found the answer in spirochete-like endocytobionts: Immunofluorescence studies would indicate homologies between microtubule-like structures in spirochetes and eukaryotic microtubules, but not between intracellular microtubules and extracellular bacterial **flagellin**. On the other hand, there are also speculations that the presumed euflagellar symbiont was photergic in nature (Schwemmler 1984). Under certain conditions, genes from such presumed photergic euflagellar symbionts which have been preserved in the host genome may be activated again. For example, certain green algae have embedded the protein-bound carotene derivate rhodopsin in their membranes (**Foster et al.** 1984). The rhodopsin, together with an adjacent eye-spot (stigma), makes light-orientated movement possible for the eukaryotic algae. Certain cells of vertebrate retina also form membrane structures in which rhodopsin is embedded. In these cases, rhodopsin serves as pigment for the visual process. On the other hand, we have already met various rhodopsins of the halophiles and one of a vertebrate in Fig. 35 (compare p. 90). According to **Foster** et al., phototaxis measurements speak in favour of homologies among the rhodopsins of vertebrates, invertebrates and unicellular organisms, and therefore of a common precursor of the rhodopsins of all eukaryotes. The amino acid sequence homologies between the halo- and bacterio-rhodopsin of halobacteria on the one hand and between the unicell- and animal-rhodopsin on the other,

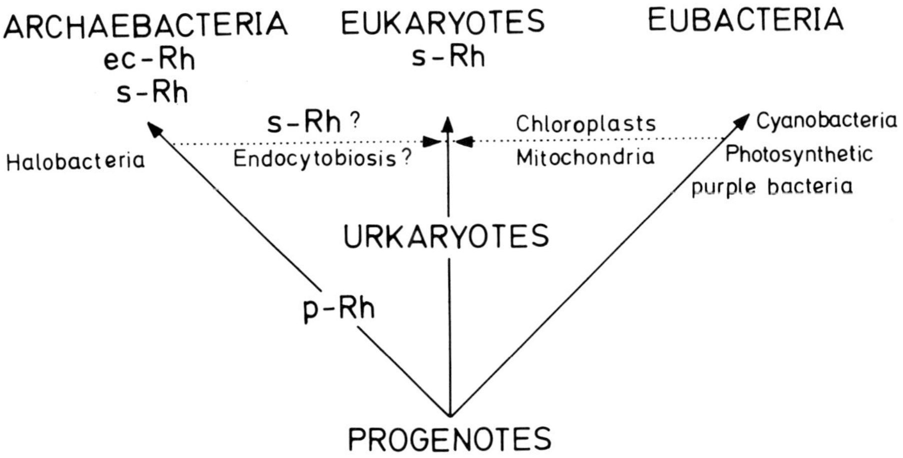

Fig. 47 a. Diagram of possible relatedness of **rhodopsins** in unicells, invertebrates and vertebrates (modified according to Dencher 1988).

ec-Rh = gene of energy converting rhodopsins

p-Rh = progenitor-rhodopsin gene

s-Rh = gene of sensory rhodopsins

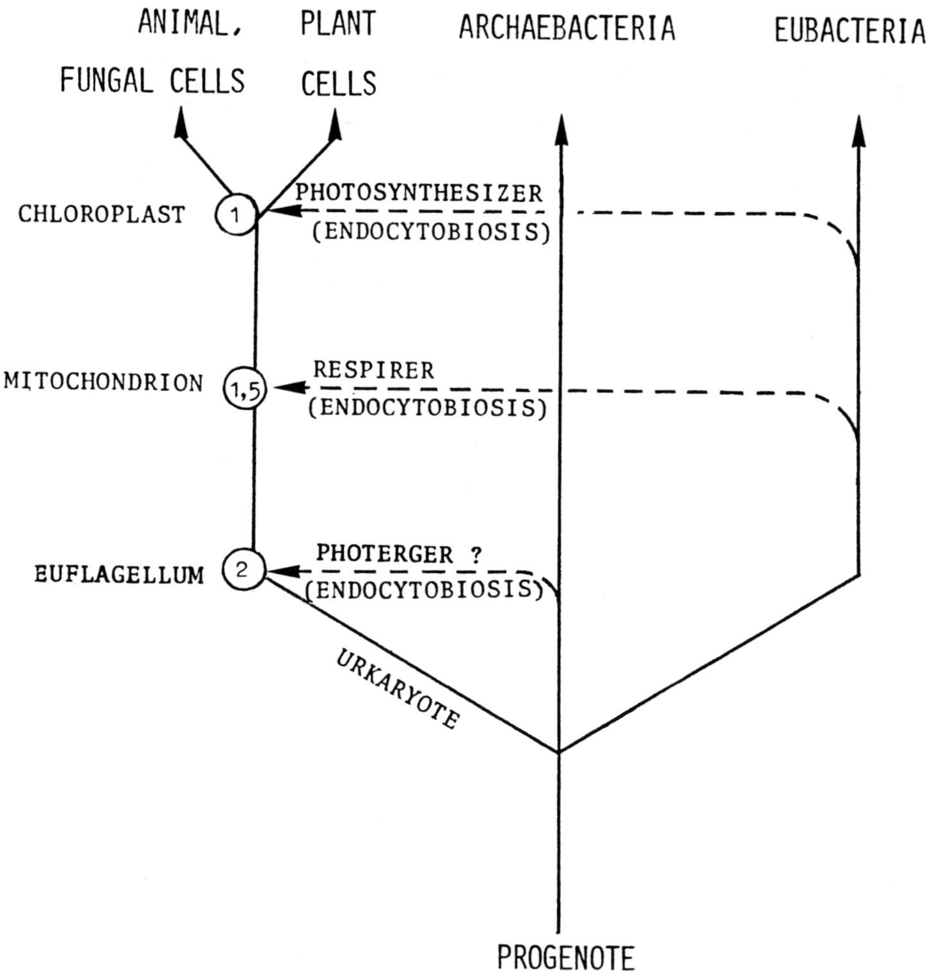

Fig. 47 b. Greatly simplified **reconstruction of eucyte evolution**. The numbers in circles indicate estimated time in billions of years (Schwemmler 1985).

however, are not highly significant (Blanck and Oesterhelt 1987). Yet, the analogous genesis can be excluded with some certainty because of the multiple similarities between both molecules, mainly in structure. Recent investigations of **Martin et al.** (1986), in this sense, support the assumptions of homologies at least between the sensory photopigments of halobacteria (compare Fig. 35, p. 94) and the eukaryotic rhodopsins. On this basis **Dencher** (1988) postulated that an archaebacterium may have produced a progenitor-rhodopsin gene (p-Rh) which later became integrated into the eukaryotic line either by gene transfer or, as **Schwemmler** (1985) postulated, by formation of endocytobiosis between an ur-karyote and a halobacterial archaebacterium (Fig. 47 a). Accordingly, it must be assumed that before the integration of the p-Rh, it split into subgroups such as those of genes of the sensory-rhodopsin (s-Rh) and of genes of the energy-converting rhodopsin (ec-Rh); only the first type of genes became integrated into the eukaryotes. Investigations in progress will soon contribute to a further understanding of the matter.

The phylogenetic age of euflagellar symbiosis, if it does exist, will also be a controversial problem. **Margulis** (1981) postulates that the mitochondrial endocytobiosis preceded the euflagellar system, while **Schwemmler** (1984) suggests that the development occurred in the reverse order, due to the loss of genetic autonomy of the euflagellar system, the partial lack of a separating membrane around the basal body, and the existence of primitive unicells (e.g. **Archamoeba**, **Metamonada**, **Mixotricha** ?; Cavalier-Smith 1987) which probably had originally no mitochondria but did have euflagella.

Another problem of eucyte evolution, for which there is still no satisfactory answer, is the **origin of the nucleus**. According to **Giesbrecht** and **Drews** (1981), the structure of the eukaryotic nucleus is not homologous to the structure of mitochondria and plastids, but is rather a composite system which evolved from more or less independent karyomers; a process which is recapitulated in the early oogenesis of many organisms. Each karyomer would ultimately derive from a simple procytic chromosomal apparatus. These considerations, of course, are speculative and ambiguous as long as there are no unequivocal experimental results. Furthermore, it is assumed that the dinoflagellates repeat early phylogenetic stages in several transitions during the development of their nuclei, in a kind of "cellular atavism". On the other hand, analyses of certain RNA sequences have revealed that the dinoflagellates on the whole do not represent an original group of organisms; in that case, they supposedly lost their histones later (Hinnebusch et al. 1981).

Recapitulating, we can conclude that the facts and inferences support the endocytobiotic nature of plastids rather than an endogenous intracellular origin, even though - in the end - this cannot be proven. They also indicate that an endocytobiotic origin of mitochondria, that is, exogenous rather than endogenous, is as good as certain. If current experiments on the eucyte motile apparatus including the euflagellum do not reveal even a trace of an exogenous genetic system - an exogenous RNA has possibly been found - both the euflagellum and other DNA-free eucyte organelles may be assumed to be of endogenous origin[1]. It is thus practically certain that the eucyte in general is not an elemental organism, but arose by assimilation, first of respiring and then of photosynthetic eubacteria, to a presumably fermenting eukaryotic host with sulfur metabolism (Fig. 47 a). Due to the lack of supporting sequence analyses, the previous integration of photergic, archaebacterial endocytobionts into a **thermoplasma-like** host to produce an euflagellated ur-karyote, remains, at least for the present, speculation. On the other hand, some eucyte organelles which lack DNA, such as dictyosomes, peroxisomes, etc., probably arose by an intracellular, i.e. endogenous development. Thus the endocytobiotic cell hypothesis has, in principle, carried the debate over the evolutionary derivation of the eucyte (more precisely: metakaryote), but, as so often happens with alternative hypotheses, the classic cell hypothesis maintains a certain validity. As it has been confirmed in many ways, one should no longer speak of the endocytobiotic cell **hypothesis**, but rather acknowledge its elevation to the rank of theory. By now, the **Endocytobiotic Cell Theory** has found its way into many textbooks. Eucyte evolution will therefore be reconstructed here on the basis of the **Endocytobiotic Cell Theory** in its extended form, which includes euflagellar symbiosis.

B. Endocytobiont model

1. Pre-eucyte

Before free oxygen was formed in the atmosphere, that is, more than $2x10^9$ years ago, the formation of endocytobioses between amoeboid procytes (similar to the archaebacterium **Thermoplasma**, which has no cell wall) and spirochete-like procytes, with a (9+2) pattern in their periplasmic cellular skeleton structure (axonema), had presumably taken place. In analogy to recent symbioses, the

---

[1] But there are mutants of chloroplasts/mitochondria without DNA. Compare also Cavalier-Smith's postulation about the symbiotic origin of DNA-free peroxisomes.

mobile procytes may have originally been episymbiotically associated with the surface of the amoeboid host. Later their head sections must have been taken up into the host cytoplasm. This would hardly have been possible by phagocytosis, as this phenomenon does not occur in procytes. It is far more likely that the symbionts would have actively penetrated the host membrane in the manner of the **Bdellovibrio** bacteria by means of lytic enzymes (e.g. Starr 1975). The membranes of the intruding head sections could then have been lysed by the host, releasing the membrane-bounded DNA of the symbiont. It might have been similar to a plasmid, which could have gradually become integrated into the host genome. The cytoplasm and ribosomes would have mingled with those of the host. The symbionts with their mobile attachment organelles may thus have come under genetic control of the host. The euflagellum, as the first typical eucyte organelle, could in principle have evolved in this fashion (Fig. 48).

Integration of the hypothetically photergic, flagellar symbiont apparently provided the host with protection from light through pigment inclusion (rhodopsin) and an increased ability to move, together with optic orientation. It was then able to live anaerobically not only in deep, dark water regions, but also in shallow, light-flooded areas. In the dark it had to live chemotrophically by fermentation with sulfur metabolism, while in the light it could live by photergy. The active motility would also have increased its radius in the search for a constantly diminishing organic substrate. This symbiosis, once formed, would have had such selective advantage that it would rapidly have spread and displaced its procyte competitors. It was a pre-eucyte in the form of a primitive unicell which, through improvement of its metabolism and motility, had "overcome" a major hurdle on the way to the later "conquest of land" by the eucytes. Today, a few of these hypothetical **euflagellated ur-karyotes** possibly still exist (e.g. rhizomastigida). They may have been repressed by the later, more extensive endocytobioses.

Fig. 48. Hypothetical formation of the most important **types of eucytes** and their organelles by endocytobiosis (adapted from Margulis 1981; for details see text).

c = chloroplast; g = euflagellum; n = nucleus; m = mitochondrion.

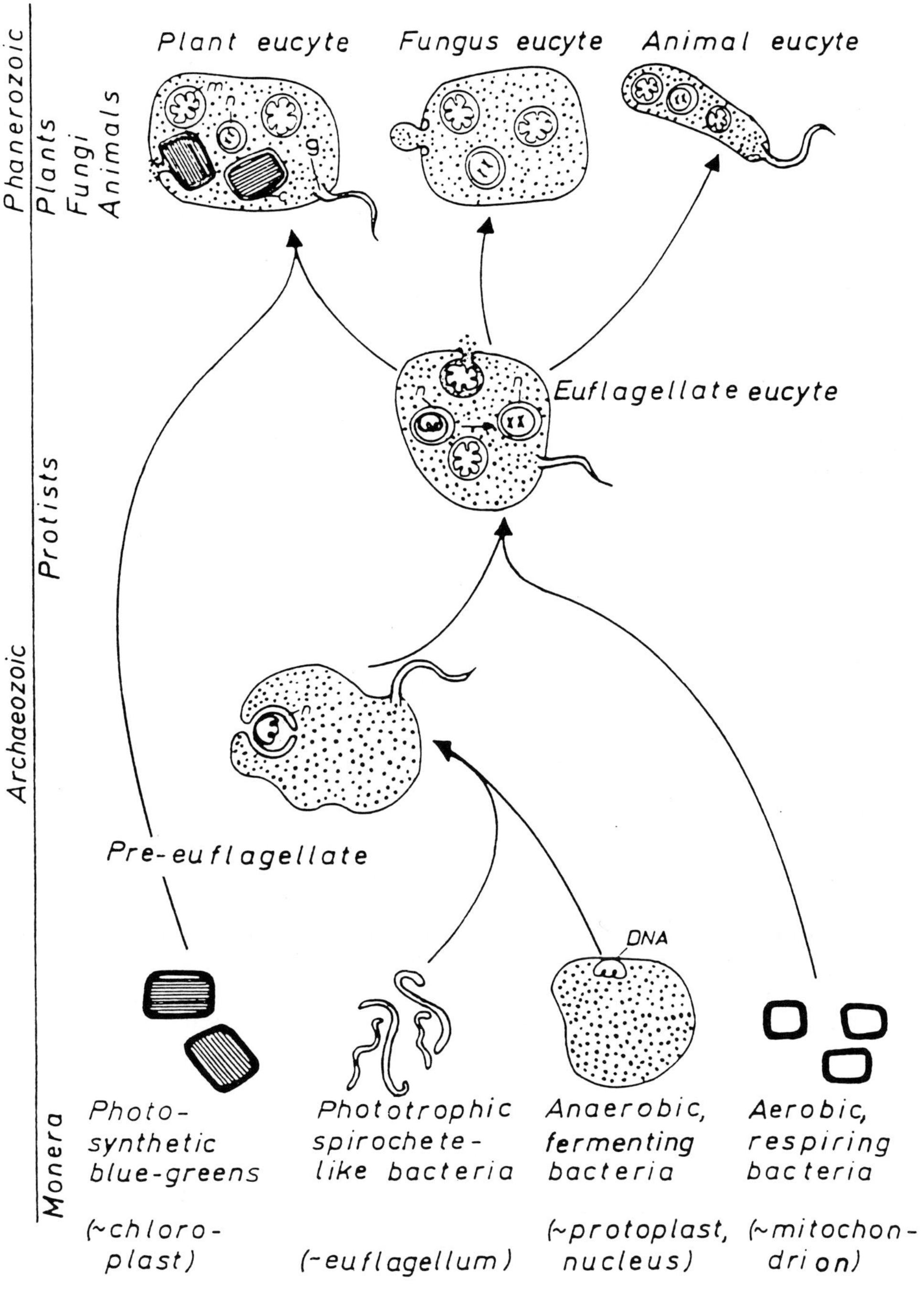
Phanerozoic
Plants
Fungi
Animals
Plant eucyte
Fungus eucyte
Animal eucyte
m
n
g
Protists
Euflagellate eucyte
n
n
Archaeozoic
Pre-euflagellate
n
DNA
Monera
Photo-
synthetic
blue-greens
(~chloro-
plast)
Phototrophic
spirochete-
like bacteria
(-euflagellum)
Anaerobic,
fermenting
bacteria
(~protoplast,
nucleus)
Aerobic,
respiring
bacteria
(~mitochon-
drion)

2. Animal and fungus cells

The second series of endocytobiotic relationships probably developed after the transition phase to the present oxidative atmosphere between $2x10^9$ and $1.5x10^9$ years ago. Presumably the pre-eucytes took up entire respiring bacteria, which must have arisen phylogenetically in the group of non-sulfur purple bacteria, and harboured them in their cytoplasm. The adoption probably occurred in exact analogy to the recent symbioses. The host membrane invaginated and phagocytosed the symbionts, along with their surrounding aqueous phase (Fig. 48). The cell wall, between the double membrane formed in this way, was gradually reduced in the course of evolution. Endocytobiont and host engaged in considerable gene exchange. On the other hand, regulation of gene exchange made it necessary for a physiological barrier to form around the host's genome. This could have taken place by invagination of the host's membrane around the site of DNA attachment, resulting in a double membrane around the host genome. As occurrs in **karyomer formation**, a species-specific number of chromosomes could then have evolved: the cell had formed a nucleus. A radical alternative to the preceding explanations is the concept that the nucleus and the endoplasmic reticulum (ER) had already formed by invagination in the ur-karyotes, preceding the formation of mitochondria, as implied in Figure 48; this concept is phylogenetically plausible, as the ER is necessary for phagocytosis. In any case, eucyte evolution is characterized by a high degree of membrane folding. The endocytobiotic phase was thus preceded and/or followed by a period of intracellular, or endogenous, differentiation. The membrane motions were apparently promoted, or even made possible in the first place, by the mobile microtubule system "imported" by the euflagellar symbiont. The most important organelles of the eucyte arose (e.g. Cavalier-Smith 1975); host nuclei and aerobic mitochondrial endocytobionts with double membranes were not the only structures formed by this motile system. Membrane folds spread like canals through the entire cytoplasm of the host cell, forming the endoplasmic reticulum. Further membrane folds in the host's cytoplasm led to the Golgi apparatus, lysosomes and the entire system of endocytosis and exocytosis[1]. The result was the separation of the various cell functions into their own reaction spaces, the compartments, which can, however, communicate with each other by membrane fusion. The increase in surface area also increased the activities of the enzymes in metabolic processes bound to membranes; this possibly affected the membrane

---

[1] **Endocytosis** is the uptake of foreign particles into tiny phagocytosis vacuoles; **exocytosis** is the extrusion of membrane vesicles filled with foreign or endogenous substances.

folds which eventually included the aerobic endocytobionts, in the case where these had not folded their membrane before phagocytosis in the host cell. These led to the formation of cristae and tubuli. Thus **mitochondrial organelles** developed from endocytobionts. A primitive eucyte had formed which may be seen as the precursor of animal cells (and possibly of some types of fungus). The mitochondrion provided the capacity for oxidative phosphorylation. In addition to the enormous increase in ATP synthesis, this meant that inorganic H^+ acceptors became available (compare Fig. 26, p. 73). This was a first step towards independence from organic substrates.

The energy supply for this cell was provided in part by inorganic substrates, which were abundantly available. Its capacity for aerobic respiration also made possible a life in the oxygen-rich regions near the surface of water. Thus the eucyte had "moved" another step closer to land. But only by an exact distribution of its chromosomes among the offspring could it reproduce precisely and thus prevail in the competition with procytes. At this point a further intracellular, endogenous differentiation occurred.

Even before the formation of the nuclear membrane, the process of differentiation of a cell division mechanism must have begun by the formation of a system of microtubules. Another microtubule system, perpendicular to the first, could have formed and provided an equatorial division furrow by membrane invagination. A primitive division mechanism may have developed which would have guaranteed that each half received a copy of the chromosome (**ur-mitosis**). After this, the nuclear membrane would have formed by invagination around the associated DNA (see above). The extra copies of the host genome (required in the absence of an exact division mechanism to insure that each filial cell received at least one copy) eventually disappeared.

The eucyte at this point of its development is represented today only by a few fungi. The differentiation of special binding sites on the chromosome itself, for attachment of the division microtubules, eventually removed the selective advantage of binding to the membrane. The centromeres of the chromosomes must have formed from these binding sites. The membrane-bounded host chromosome with its transcription and translation system developed more rapidly and differently from the membrane-bound mitochondrial chromosomes (cf. 80S, 70S ribosomes). By repeated duplication, the host chromosome may have grown rapidly to a multiple of its original length (Haapala and Sorsa 1976). The recent euflagellated archamoeba may be a descendent of such eucytes with chromosomes and ur-mitosis characteristic for these species. In the course of further evolution, the mitotic division mechanism improved. Homologues of the euflagellar basal body with the pattern (9+0) differentiated to centrioles. One centriole pair moved to opposite poles of the cell. By formation of a mitotic spindle and dissolution of the nuclear membrane they achieved the separation of the replicated chromosomes. Nuclear division was followed by division of the cell. **Mitosis**, or eumitosis, had developed.

The sexual fusion of two eucytes, possibly as a response to starvation or recombination, could then have led to the evolution of **meiosis**. Meiosis requires mechanisms allowing an exact pairing of homologous chromosomes.

3. Plant cell

The third endocytobiosis formation in eucyte evolution may then have occurred about 10^9 years ago (Fig. 48). Some eucytes must have incorporated photosynthetic bluegreen bacteria in phagocytosis vacuoles (Schnepf and Brown 1971). As in recent endocytobiosis, the cell wall of the symbiont would have been reduced, until the organelle was surrounded only by a double membrane (or triple membrane as in protists). Thus it could, like the mitochondria, exchange genetic material with the nucleus. This led to a loss of genes from the endocytobiont genome, and to the formation of the **plastid-like organelles** of the phytoflagellates. The plastids provided the eucyte with complete independence from organic substrates, and achieved the definitive precondition for the conquest of land. However, the settlement of land was first achieved by multicellular systems (compare p. 139).

V. Summary

About $2x10^9$ years ago, the evolution of the most important procyte metabolic types had in principle been completed. Eventually the evolution of procytes may have reached a dead end. Different types of metabolism within a single procyte, such as fermentation, photergism, respiration and photosynthesis could have been fully active only in succession, never simultaneously. These metabolic processes, which in part counteract each other, would have inhibited each other because there was no subdivision of the cell into separate reaction spaces (**compartments**). It was discovered by **Pasteur** that anaerobic fermentation and aerobic respiration in bacteria cannot occur at the same time (Pasteur effect). The conversion from one metabolic type to the other is associated in modern procytes with internal rearrangements which may be profound. For example,

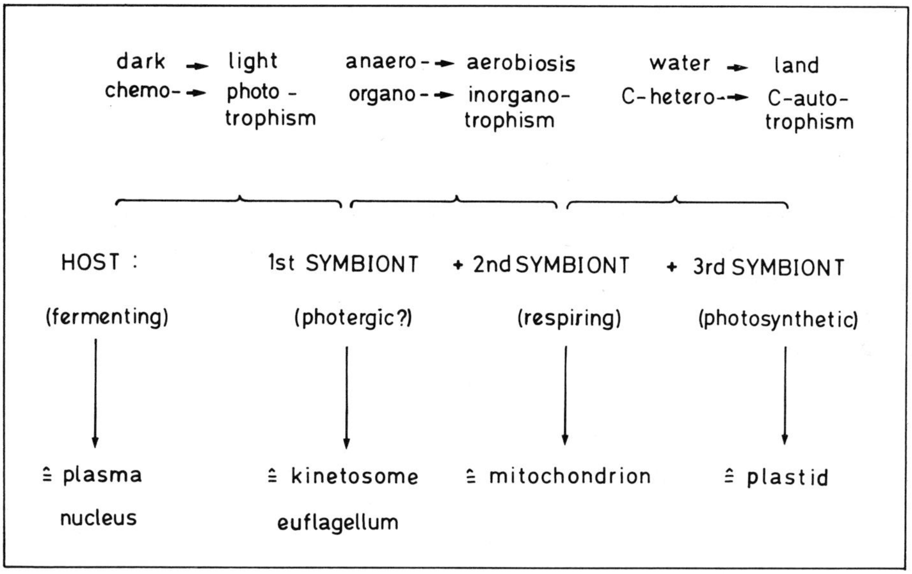

Fig. 49. Origin of the DNA- and RNA-containing organelles as a result of endocytobiosis (**Endocytobiotic Cell Theory**) in connection with changes in biotope and ecological niche.

when certain bacteria convert from respiration to photosynthesis, they must form thylakoids by invagination of the plasma membrane and embedding of photopigments. Thus a certain amount of time is required before a newly turned-on metabolic pathway is completely functional. It follows that about $2x10^9$ years ago there would have been selective pressure leading to compartmentalization of different metabolic pathways, in order to make them available simultaneously and immediately. Compartmentalization evolved in a much shorter time via the formation of intracellular symbioses than it would have by intracellular differentiation. This is because there were prefabricated parts available for the formation of endocytobioses, and these were more quickly incorporated than adequate intracellular reaction spaces could have evolved. Thus almost inevitably, although it happened by chance, a series of mutually interactive endocytobioses arose. The results were endocytobiotic complexes which united important types of metabolism: fermentation, photergism, respiration and photosynthesis. Such complexes were then in a position to immediately and simultaneously occupy each of the four elementary biotopes which had been previously conquered by the procytes (Fig. 49). The incorporation of photergers into fermenters would have enabled the change from dark to light and from chemo- to phototrophism; the incorporation of respirers the change from anaerobiosis to aerobiosis and from organo- to inorganotrophism (inorganic H^+ acceptors); finally, the incorporation of photosynthesizers made possible the transition from water to land (air) or from C-heterotrophism to C-autotrophism.

It must be assumed that the process of integration of bacterial endocytobionts as DNA-containing organelles in the eucyte is far from complete. Rather, the process today appears to still be in full swing. Thus the endocytobiotic evolution of the eucyte was and still is achieved by formation of new obligate endocytobioses between eukaryotic host cells and bacteria which restrict themselves to that particular host.

VI. Micro-mechanisms of biogenesis

In the reproduction of precyte, procyte or eucyte by division, the filial cells could not have been absolutely identical in every case, since this would have prevented progress in evolution. Evolution occurs only when the environment selects better adapted offspring through the mechanism of reproductive success; in time, the less successful forms are displaced. This, however, requires that the transmission of genetic information should be subject to larger or smaller genetic changes (**mutations**), which are passed on to the offspring. Point mutations are those changes for instance which occur when an incorrect nucleotide is incorporated into DNA during replication. When the genetic information is translated into protein, a different amino acid may be incorporated as a result of the change in nucleotide sequence. Another form of mutation involves entire chromosomes or even sets of chromosomes when genetic material is lost or added. These mutations are called chromosomal or genome mutations. Mutation and **selection**, as the "motor of evolution", drive the development from simple to complex.

Charles Darwin (1859) was the first to recognize and describe this micro-mechanism of biological evolution. According to his theory, mutations arise by chance and without direction, and are then inherited by the offspring, leading to genetic variation among them. The theory states that those genetic variants which are best adapted to their external and internal environments are most likely to reproduce.

There are a number of micro-mechanisms of bioevolution. For example, if two individuals mate and reproduce successfully, the genetic material of the two is mixed, a process called **recombination**. Like mutation, it provides raw material for evolution. If a group of individuals separates from its population, so that gene exchange no longer occurs with the original population, the isolated group may change in such a way that when it later comes into contact again with its home population, individuals from the two groups may no longer be able to reproduce fruitfully (genetic drift). In this case, a new species has arisen. Thus **isolation and genetic drift** is another important micro-mechanism in the origin of new species.

VII. References

Altmann, R. (1890): *Die Elementarorganismen und ihre Beziehung zu den Zellen*. Viet and Comp., Leipzig.

Arnold, C.G. (1977): *IV. Extranuclear Heredity: The phylogenetic origin of extranuclear heredity*. Fortschr. Bot. 39 182-191.

Blanck, A. and Oesterhelt, D. (1987): *The halo-opsin gene. II: Sequence, primary structure of halorhodopsin and comparison with bacteriorhodopsin*. EMBO J. 6 265-273.

Cavalier-Smith, T. (1975): *The origin of nuclei and of eukaryotic cells (a review).* Nature 256 463-468.

- (1987a): *The origin of eukaryote and archaebacterial cells.* Ann. N.Y. Acad. 503 17-54.

- (1987b): *The simultaneous symbiotic origin of mitochondria, chloroplasts and microbodies.* Ann. N.Y. Acad. 503 55-71.

Darwin, C. (1859): *On the origin of species by means of natural selection, or the preservation of favoured races in the struggle for life.* John Murray, London.

Dencher, N.A. (1988): *Rhodopsin-like membrane proteins in the archaebacterium* Halobacterium halobium*: A clue to cellular evolution*; Endocyt. C. Res. 5 (1) 1-16.

Doolittle, W.F., Bonen, L. (1981): *Molecular sequence data indicating an endosymbiotic origin for plastids.* In: Fredrick, J. (ed.): Origins and evolution of eukaryotic intracellular organelles. Ann. N.Y. Acad. Sci. 361 248-256.

Foster, K.W., Saranak, J., Patel, N., Zarilli, G., Okabe, M., Kline, T. and Nakanishi, K. (1984): *A rhodopsin is the functional photoreceptor for phototaxis in the unicellular eukaryote* Chlamydomonas. Nature 311 756-759.

Giesbrecht, P. and Drews, G. (1981): *Die "Kernstrukturen" der Bakterien und ihre Beziehungen zu denen der "Mesokaryoten".* In: Metzner, H. (ed.): Die Zelle: Struktur und Funktion. Wiss. Verlagsges. Stuttgart; 3rd edition.

Görtz, H.D. (1980): *Nucleus-specific symbionts in* Paramecium caudatum. In: Schwemmler, W. and Schenk, H. (eds.): Endocytobiology I, 381-392. Walter de Gruyter Verlag Berlin, New York.

Grimstone, A.V. (1964): *The structure of* Mixotricha *and its associated microorganisms.* Proc. Br. Royal Soc. (London) 159 668-686.

Haapala, O. and Sorsa, V. (1976): *Evolution of eukaryotic chromosome organisation.* Biol. Zbl. 95 317-325.

Hartmann, H. (1975): *The centriole and the cell.* J. theoret. Biol. 51 501-9.

Hinnebusch, A.G., Klotz, L.C., Blaken, R.L. and Loeblich, A.R. (1981): *An evaluation of the phylogenetic position of the dinoflagellates.* J. mol. Evol. 17 334-347.

Kaplan, W. (1978): *Der Ursprung des Lebens. Biogenetik, ein Forschungsgebiet heutiger Naturwissenschaft.* 2nd edition; Georg Thieme Verlag Stuttgart.

Louis, C. and Laporte, M. (1969): *Charactères ultrastructureaux et différenciation des formes migratrices des symbiotes chez* Euscelis plebejus *(Homoptera, Jassidae).* Ann. Soc. Ent. 5 799-809.

Mahler, H.R. and Raff, R.A. (1972): *The non-symbiotic origin of mitochondria.* Science 177 575-582.

Margulis, L. (1970): *Origin of eukaryotic cells.* Yale University Press New Haven, London.

- (1975): *The microbes contribution to evolution.* Biosystems 7 266-292.
- (1981): *Symbiosis in Cell Evolution.* W.H.Freeman San Francisco.

Margulis, L., To, L. and Chase, D. (1978): *Microtubules in prokaryotes.* Science 200 1118-1124.

Martin, R.L., Wood, C., Baehr, W. and Applebury, M.L. (1986): *Visual Pigment Homologies Revealed by DNA Hybridization.* Science 232 1266-1269.

Müller, H.J. (1949): *Zur Systematik und Phylogenie der Zikaden-Endosymbionten.* Biol. Zbl. 68 343-368.

Pickett-Heaps, J. (1974): *The evolution of mitosis and the eukaryotic condition.* Bio System 6 37-48.

Schenk, H.E.A. (1977): *Inwieweit können biochemische Untersuchungen der Endocyanosen zur Klärung der Plastiden-Entstehung beitragen?* Arch. Protistenk. 119 274-300.

Schimper, A.F.W. (1883): *Über die Entwicklung der Chlorophyllkörner und Farbkörper (1. Teil).* Bot. Z. 41 105-114.

Schnepf, E. and Brown, R.M. (1971): *On the relationship between endosymbiosis and the origin of plastids and mitochondria.* In: Reinert, J. and Ursprung, H. (eds.): Results and problems in cell differentiation, Vol. 2, 299-322; Springer Verlag Heidelberg, London, New York.

Schopf, J.W. and Zeller-Oehler, D. (1976): *How old are the eukaryotes?* Science 193 47-49.

Schwemmler, W. (1984): *Reconstruction of cell evolution: A periodic system.* CRC Press Boca Raton, Florida.

- (1985): *Endocytobiosis formation: macromechanism of cell evolution.* Endocyt. C. Res. 2 191-211.

Starr, M. (1975): *The bdellovibrio association as model.* In: Jennings, D. (ed.): Society of experimental biology, symposion 29., University Press Cambridge.

Taylor, F.J.R. (1974): *Implications and extensions of the serial endosymbiosis theory of the origin of eukaryotes.* Taxon 23 229-258.

Whatley, J.M. (1976): *Bacteria and nuclei in* Pelomyxa palustris*: Comments on the theory of serial endosymbiosis.* New Phytol. 76 111-120.

Woese, C.R. (1981): *Archaebacteria.* Scientific American 244 (6) 94-106.

Woese, C.R. and Fox, G.E. (1977): *The concept of cellular evolution.* J. mol. Evol. 10 1-6.

Chapter 5 Sociogenesis

I. Phenomenon.

The evolutionary processes discussed so far occurred mainly in a water environment. A number of difficulties had to be overcome before life could be established on land, e.g. the danger of drying out, the large changes in temperature, and the problems of gravity, absorption of nutrients and water, and reproduction. Green-algae-like plants probably settled in terrestrial biotopes before animals could (Kuhn and Probst, 1980). They developed special tissues to cope with the problems of life on land: external tissues prevent drying out, and other parts of plants influence their temperature when it is either too cold or too hot. Support tissues overcome the gravitational force of the earth, while other cellular structures provide a constant supply of water and nutrients or meet the special requirements of terrestrial reproduction.

Such multicellular or **polycytic systems** which developed in plants and animals first in water and then on land, all have a characteristic hierarchic structure. They consist of diversely specialized systems of organs which are associated in one integrated **organism** (Fig. 50); each **organ system** consists of different **organs** which in turn consist of different types of **tissue**, and each type of tissue is made up of **eucytes** of a single type. Thus, the animals have developed many specialized organs, e.g. integument, skeleton, musculature, nerve system, blood vessels, digestive tract, respiratory, excretory and reproductive systems, which are well integrated into a functional organism.

The further development of a part of the nervous system finally led to the evolution of the human brain and consciousness, which induced the new, cultural phase of evolution.

The process by which eucytes join into organized structures is sometimes called cellular "association". At this simple level of so-called "sociogenesis", communication between individuals seems to be exclusively chemotactical, that is, it occurs by means of chemical attractants. For example, the attraction of "social" amoeba to form the fruiting body of the slime mould **Dictyostelium discoideum** seems to be effected by excretion of an organic substance (cyclic adenosine monophosphate) which is formed as a result of starvation (reviewed by Sengbusch 1979). The evolution of communication requires simultaneous development of organs sending out chemical signals and of receptors able to process the encoded information. In colonies of social insects, communication mechanisms using mechanical , optical and chemical signals work together. In associations with still more highly evolved mechanisms of communication, such as those of the apes, even more complex forms have developed, such as rituals and traditions. Among human beings, speech, writing, laws and conventions have developed to regulate the contact of members of a society or between different societies.

It should be stated, to prevent misunderstanding, that the following discussion does not concern the details of various mechanisms of communication within associations. Although this is undoubtedly an attractive subject, it is not within the purview of this book; for the anthropological side of sociogenesis (development of culture, sociology, psychology, etc.) the reader is referred to the voluminous special literature (e.g. Childe 1975, Leakey and Lewin 1978, Eder 1980, Kull 1979, Riedl 1980). **Instead, in this chapter, the biological roots of human sociogenesis are exposed,** i.e. the evolution of the biological structures which are a necessary prerequisite for human sociogenesis. They include elementary processes which have led to the transition from eucyte to polycyte and on to the human organism and consciousness and to achievements resulting from consciousness.

II. Hypotheses

The current concept of the pathway from eucyte to polycyte and on to modern complex organisms (plants, fungi and animals, including human beings) goes back to **Ernst Haeckel** (1834-1919). He stated: "The organic individual repeats, during the rapid and short course of its individual development, the most important of those changes in form which its ancestors went through in the slow and long course of their paleontological development." This statement was then shortened to the basic **rule of biogenesis**: "Ontogeny (individual development) is a short and rapid recapitulation (repetition) of phylogeny." For example, all higher vertebrates pass through a "fish stage" in their ontogenetic development (embryonic anlage for gill arches etc.), like stages in the evolution of their aquatic ancestors.

It was soon realized, however, that the biogenetic rule, in the above general form, does not apply to all cases. **Portmann** remarked in 1969 that "**the embryonic stages of higher vertebrates display some traits only temporarily, which persist in fish in their completed form.**" The modern formulation of the biogenetic rule

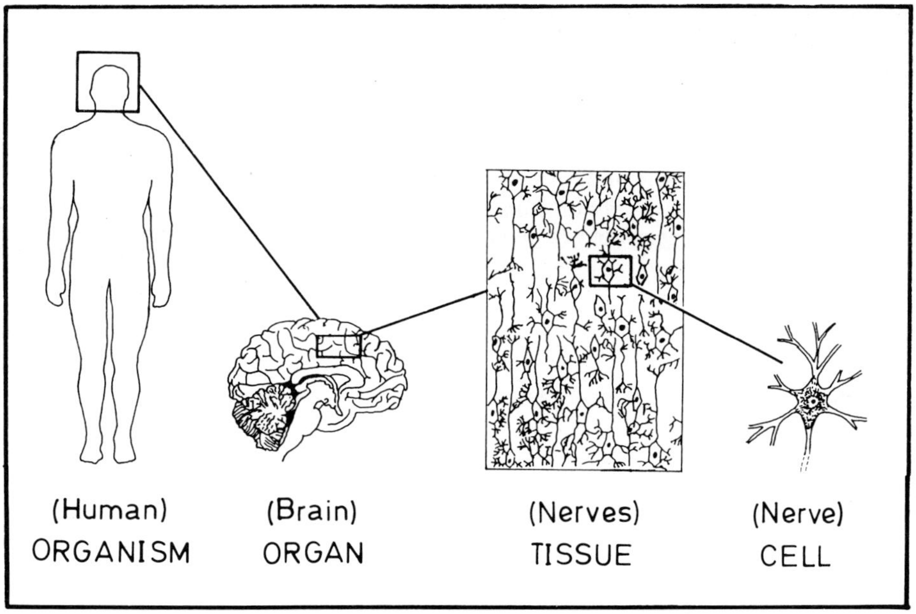

Fig 50. Polycytic organisms are made up of organ systems, as shown here in the case of the **human brain**. Each organ system consists of different organs each of which is a functional unit. Organs are made up of various types of tissue, with each type of tissue consisting of a single type of cell. The knowledge that all organisms consist of cells goes back to the classic cell concept of Schleiden and Schwann (1839; see historical review in Denffer et al., 1983).

thus acknowledges this limitation: ontogeny is to a certain degree a short and rapid recapitulation of certain embryonic stages in phylogeny.

The hypothesis proposed by Haeckel on the origin of human beings as multicellular organisms has now been confirmed by a large body of scientific evidence. The hypothesis is supported by ontogeny (embryology), phylogeny based on homologies and fossils, brain research (neurobiology), speech research (part of ethology), and by sequence homologies of proteins, DNA/RNA, and chromosomal structures.

A radical alternative hypothesis is the idea of independent polycytes evolving first to various types of tissue and/or organs and later combining with each other, in a kind of symbiosis to form a multifunctional organism. In the following, the recent data will be presented to evaluate which one of the hypotheses will be supported (Fig. 51).

III. Data

A. Embryonic development

The first phase of **ontogeny** is embryogenesis, the development of the fertilized egg or vegetative generative cell into a viable organism. During this process, polycytic organisms pass through developmental stages which are in part analogous, but which in their details demonstrate a high degree of variability. At the beginning of the development of every polycyte is a single eucyte. In sexual reproduction, it is the fertilized egg (**zygote**); in asexual reproduction, the vegetative generative cell. Highly developed polycytic systems can serve as an example of typical ontogeny common to higher animals (e.g. humans; Fig 52). Repeated divisions or cleavages of the zygote lead to a solid sphere of undifferentiated cells, the **morula**. The morula develops into a **blastula** as the embryonic cells arrange themselves around the periphery and leave a fluid-filled cavity in the centre. Next, the process of gastrulation transforms the embryonic cells into the three primary germ layers; the outermost layer is the **ectoderm** and the innermost layer is the **endoderm**; in the further course of this development a third germ layer, the **mesoderm**, develops. Concerning the diverging development of the germ layers in mammals or humans (trophoblast / embryoblast), see specific literature (e.g. Romer 1986).

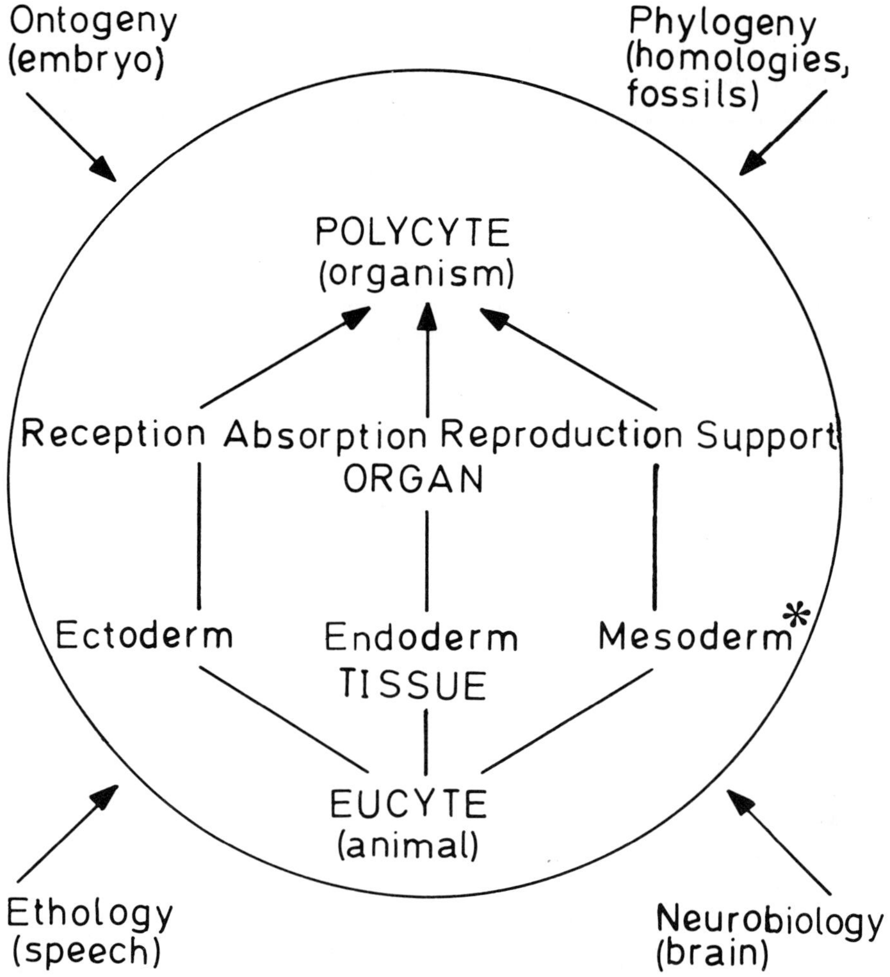

Fig. 51. Research fields contributing to the study of the evolution of animal polycytes, including humans.

* The mesoderm in different groups of animals differs with respect to its embryogenetic origin, and its definite structure and function.

The embryo is roughly spherical up to this stage, then it begins to elongate. Along its back, in the ectoderm, the process of neurulation (**neurula stage**) starts during which the neural tube forms, which is the precursor to the central nervous system including the brain; other organ systems which later develop from the ectoderm include the sensory cells, the lens of the eye and the epidermis. The mesoderm gives rise to an elastic rod along the neural tube, the chorda; this is later supplemented by the vertebral column. This feature gives the vertebrates (fish, amphibians, reptiles, birds, mammals) their name. In addition to the vertebral column, the mesoderm produces bones, cartilage, muscles, lymph and blood vessels with heart, spleen, kidneys and gonads. The endoderm gives rise to the epithelia of the lungs, stomach, intestine, bladder, pancreas, liver and to thyroid and thymus. The following scheme summarizes the essential steps of embryogenesis; it adequately depicts the origin of the hierarchic order in a polycytic organism.

Zygote --> Morula--> Blastula--> Gastrula--> Neurula--> Human
Cell --> | <--- Tissue ---> |<- Organ -> | <- Organism

Haeckel's biogenetic rule states that the basic stages of phylogeny occurred in a similar pattern, as we shall investigate more closely in the following.

B. Phylogenesis

Embryogenesis begins with the fusion of an egg cell with an **euflagellated sperm cell**. This is suggestive - in **Haeckel**'s sense - of a phylogenetic parallel to the structure of euflagellated algae (flagellates) which probably evolved between $1.5x10^9$ and 10^9 years ago (Fig. 53; compare p. 128). The fusion of sperm and egg cells into multicellular organisms is also prefigured in unicellular organisms (Hoff and Miram 1979). In the process of the **fusion of two flagellate** cells with subsequent mixing of their genetic information (both nuclear and cytoplasmic), we have the beginnings of sexuality.

The fertilized egg cell, the **zygote**, forms the morula by repeated synchronous cell divisions, so-called cleavages. The morula consists of identical, non-differentiated cells which cling together. On the phylogenetic side, the corresponding structure is seen in the **colonies of some algae** which are formed by cells which have not separated after division. Such colonies may have first formed 10^9 years ago.

The **blastula** is preserved as the final developmental stage of only one living genus of organisms, the green alga **Volvox**, which forms a single-layered sphere

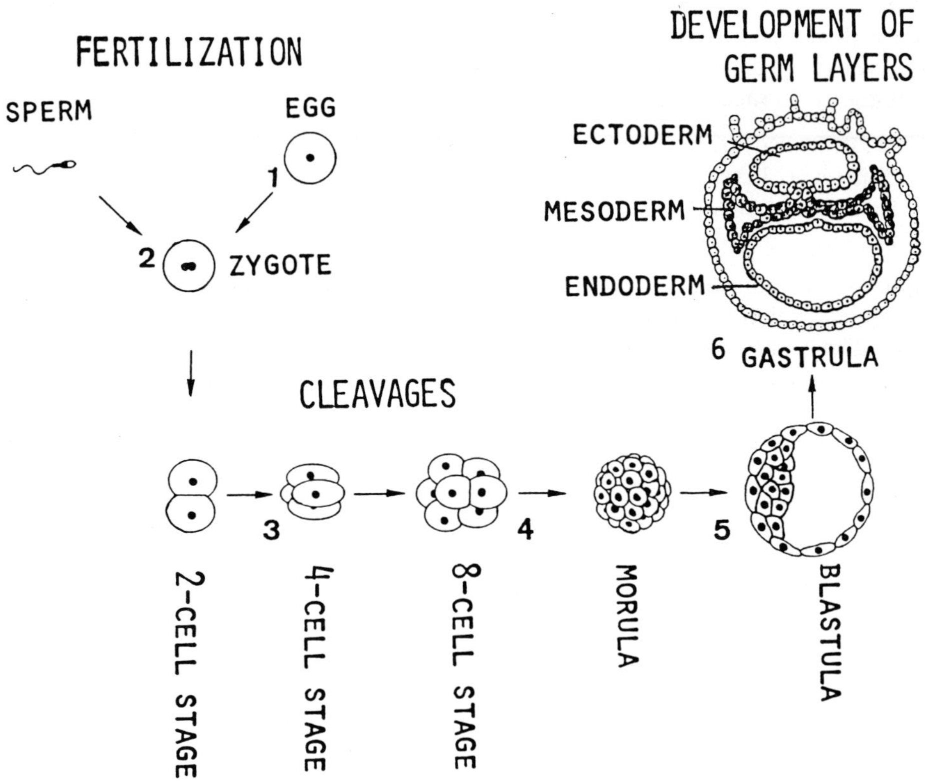

Fig. 52. After fertilization of a human egg cell (1) by a sperm, mitotic divisions of the zygote (2) lead to the morula, which migrates through the oviduct into the uterus (3,4) and embeds itself there (5). Further mitoses, cell displacements and rearrangements convert the morula into the blastula (humans: blastocyst, 5) and then the three embryonic layers form: ectoderm, endoderm and mesoderm (6). The embryonic germ layers develop into the various parts of the body and into the organs (for details, see Romer 1986, in particular the diverging development of the germ layers in mammals and humans: trophoblast/embryoblast).

of up to 20,000 euflagellated cells. Nutrition and information can be exchanged between these cells. New spheres form in the hollow centre through cell division; once these have reached a certain size, the primary sphere bursts and dies after the young spheres have been released. **Volvox** has a structure which is remotely similar to the blastula.

The structural traits of the sack-shaped **gastrula** of the coelenterates are reminescent. Like the young gastrula, the **coelenterates** consist of two layers, ectoderm and endoderm. They are thought to have developed about 700 million years ago (see also the placula theory and **Trichoplax**: Grell, Biuz 4 65-71, 1974).

The **neurula** stage corresponds to the evolution of the **chordates** and vertebrates, which according to the fossil record must have appeared about 500 million years ago. The embryonic stages with the anlage for the extremities recapitulate the formation of the first organisms with fore- and hind-legs, which has been dated about 350 million years ago. Finally, the formation of the typical mammalian habitus (e.g. anlage of mammae and fur) in the embryo can be related to the appearance of the first mammals about 200 million years ago.

Now that the transition from eucyte to polycyte has been explained, at least in rough outline through use of the biogenetic rule, it is interesting to observe more closely the development of the organ on which the subsequent cultural phase of evolution is based: the vertebrate brain.

C. Brain evolution

In illustrating the structural and functional assignments of the brain's diverse regions, it may be useful to first describe the **ontogenesis of a typical vertebrate brain**. According to comparative investigations of **neurobiology** the brain of the vertebrate embryo develops more rapidly than most of the other organs (Langman 1977, Romer 1986). In early embryonic stages the brain merely consists of an enlargement of the dorsal neural tube in the head region. In a developing embryo, the brain soon expands forming two vesicles: the anterior, slightly bent forebrain (prosencephalon) and the hindbrain (rhombencephalon). The forebrain becomes differentiated into the anterior cerebrum (telencephalon) and the interbrain (diencephalon, gives rise to the thalamus and hypothalamus; cp. Fig. 54). The rhombencephalon produces the midbrain (mesencephalon), the metencephalon (dorsal part becomes the cerebellum) and the myelencephalon (medulla oblongata) reaching into the spinal chord. Towards the end of ontogenesis the telencephalon develops into the cerebral hemispheres and the paired olfactory lobes.

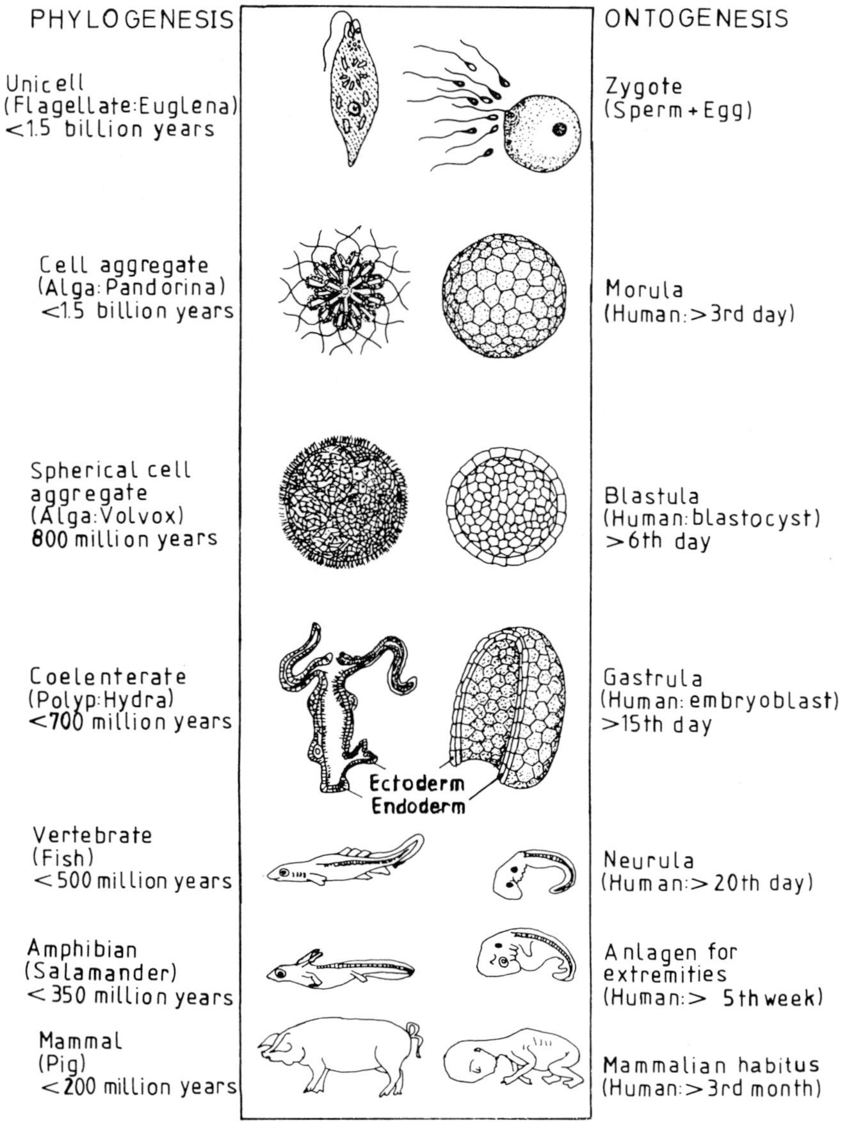

Fig. 53. Parallels in the evolution from unicellular to multicellular organisms in **embryogenesis** and **phylogenesis** (partially adapted from Hoff and Miram 1979).

The **phylogenesis of the brain** may be reconstructed by comparing species possessing a stage of original brain anatomy (Kahle 1986). Accordingly, the telencephalon of amphibians and reptiles, appears as a minor part of the well developed olfactory lobes (Fig. 54). In fish (chondrichthyes, osteichthyes), amphibians and reptiles, the di- and mesencephalon are not covered by the enlarging hemispheres, as they are in birds and mammals. In original mammals the telencephalon begins to expand, covering other parts of the brain, and beginning in lemurs, the cerebrum completely surrounds the di- and mesencephalon. In birds and mammals the elaboration of the cerebrum and the cerebellum is linked to the development of a new structure, the pons, a mass of fibres connecting these two areas and enabling a direct exchange of information. In the course of the phylogenetic development of the vertebrate brain the cerebrum (telencephalon) as well as the interconnections of different regions of the brain enlarge permanently.

A **comparison of the brain's ontogenesis with its phylogenesis** shows that its different regions develop at different times; this is correlated with their diverse functions. Those parts of the human brain which develop early are already quite elaborate in primitive vertebrates, and control elementary instinctive functions of life such as breathing, thermoregulation, thirst and appetite etc.. These regions of the vertebrate brain, except the cerebrum (telencephalon) and the cerebellum (dorsal region of the metencephalon), are collectively called the **brainstem**. During a late phase of ontogenesis, and only in the phylogenetically younger vertebrates (birds, mammals), the cerebellum and cerebrum, the centres of coordination and correlation, become convoluted. All the instinctive and learned patterns of motor activity are stored in the cerebellum, which is controlled directly by the cerebrum and coordinates the complicated orientation and stabilization in three-dimensional space. Thus the cerebellum of birds (a particularly large area, the neocerebellum, is concerned with flight) and of **Homo sapiens** (upright gait) is well developed and convoluted, whereas in the less agile reptiles the cerebellum is only moderately developed (Fig. 54). The **cerebrum** is the centre for conscious processing of information and directly controls the cerebellum. We now have at our disposal **topographic mappings** of the telencephalon which depict its functionally defined areas (Fig. 55). Some areas receive sensory stimuli from the different receptors, and other centres send out motor stimuli to the musculature etc..

The original vertebrate cerebrum served a rather narrow purpose, chiefly the reception of olfactory/chemical stimuli (Hoff and Miram 1979). In the lower vertebrates it is therefore relatively small, compared to the brainstem (Fig. 54). In fish, amphibians and reptiles, instinctive behaviour dominates, coordinated by the brainstem. In birds and mammals the telencephalon has evolved further, increasing considerably in volume and becoming convoluted at the surface. In the course of vertebrate evolution this increase in surface is linked with the rapidly

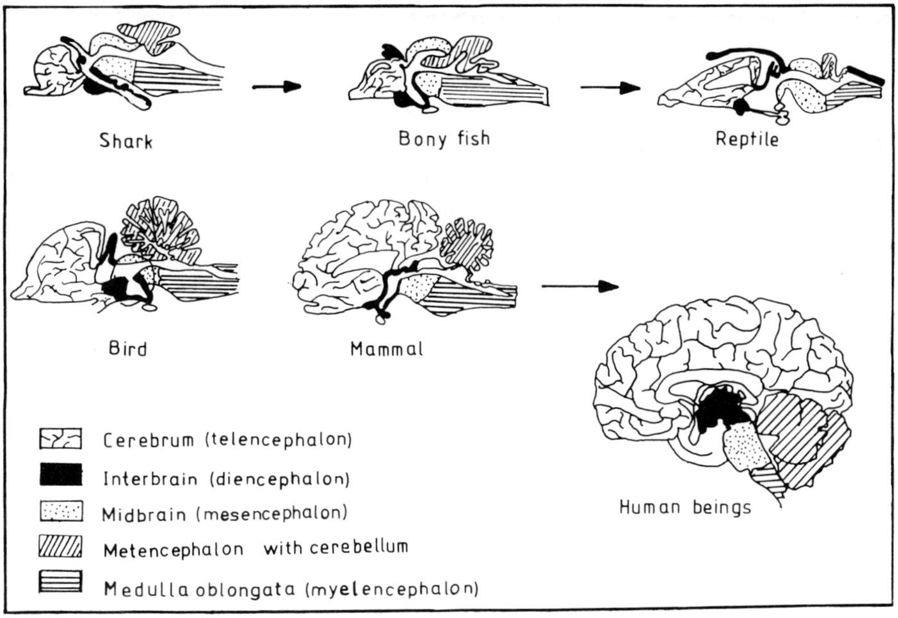

Fig. 54. Schematized evolution of the vertebrate brain from fish to human being. Drawings of the individual brains are not to the same scale (adapted from Hoff and Miram 1979).

increasing number of nerve cells (neurons) with ever more interconnections. The human cerebrum contains about $1.5x10^{10}$ nerve cells. The number of neurons in the brain is directly proportional to its capacity. Birds and mammals dispose of an enormously enlarged repertoire of behaviour, linked with their progressive capacity for learning. The main phase of imprinting and learning in most birds and mammals is limited to their juvenile period, whereas the human capacity for learning persists a lifetime. **Homo sapiens** has the most elaborate cerebrum of all creatures on this planet. In contrast to other vertebrates the human telencephalon bears an enormous **learning capacity** and **memory** as well as **intelligence, self-consciousness** and a **speech centre**. About 3 million years ago, this cerebral capacity initiated the new phase of cultural evolution, combined with structural changes in habitus and incisive alterations of the behaviour pattern of humankind.

An essential element of the complicated interactions of this process is the evolution of speech, which will therefore be treated in the following.

D. Evolution of speech

The highly developed cerebrum together with the brainstem permits birds and many mammals (especially whales) to engage in **acoustic communication**. Comparative study of behaviour, or **ethology**, has discovered many examples of this (for a review, see Hoff and Miram 1979). They range from simple communication mechanisms in mating and territorial defence systems to complex word-like systems in social bands. All of these mechanisms have in common that they are highly instinctive, and can vary only to a slight degree. Even the highly developed apes are not able to formulate words, in spite of intensive training. Human beings are able to articulate abstract speech in acoustic symbols. **Speech is one of the most important species characteristics of the human mind and enables us to model the real world in words and sentences independently of space and time.** It is a historically developed and integrated wholeness consisting of a differentiated vocabulary, parts of sentences and of basic forms of syntax. Speech evolves as a collective means of communication and the individual develops it only in contact with other persons. The structural peculiarities of the human larynx, buccal cavity and throat make speech possible (Fig. 56). These speech organs, together with a developed musculature, make possible the wide spectrum of spoken sounds. In addition, the human cerebrum contains two highly developed speech centres (cp. Fig. 55, Nr. 2-4,6), one for understanding and the other for the motor aspects of speech. The latter is completely absent in the

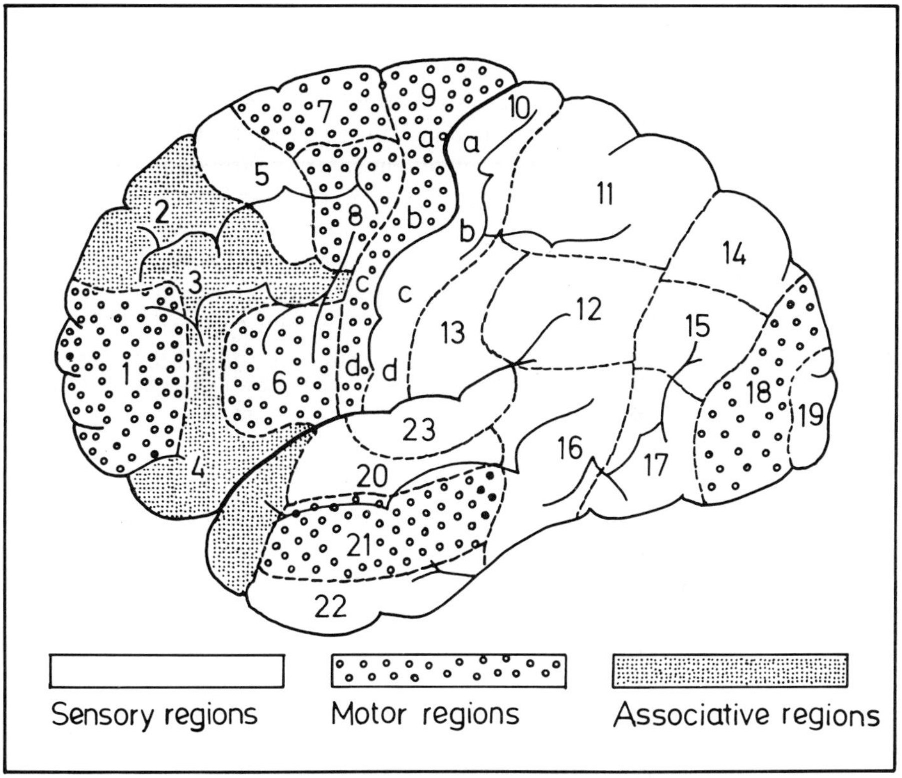

Fig. 55: Regions and functions of the left hemisphere of the human cortex (telencephalon). (According to Hoff and Miram 1979)

1.	motor sequences	2.	impetus	3.	creative thinking
4.	frame of mind	5.	sensation of orientation	6.	motor speech centre
7.	skills	8.	writing centre		
9.	locomotion a) leg, b) trunk, c) arm, d) face				
10.	sensation of pain, temperature, contact			11.	independent actions
12.	centre of activity	13.	tactile memory	14.	memory of locality
15.	reading centre	16.	understanding of concepts		
17.	colour recognition and locomotion			18.	eye movement
19.	sight	20.	hearing	21.	acoustic orientation
22.	recognition of music and sound			23.	taste

closely related apes, although their cerebrum shows the first beginnings of a centre for understanding speech.

Since the ancestors of the recent **Homo sapiens** are extinct, the evolution of the motor speech centre may only be analysed using moulds of fossil crania. Like other regions of the cerebrum, the human motor speech centre has a typically convoluted surface, its pattern causing complementary impressions on the inside of the skull during the simultaneous development of brain and skull. According to the theory by **H. Spatz**, the regions of the brain, undergoing a progressive evolution, will leave exceptionally clear impressions (in Kahle 1986). Such impressions of fossil brains indicate that even the earliest pre-human[1] (**Homo habilis** = tool-using man) had the beginnings of a motor speech centre. Even though this does not permit us to infer abstract speech, it does suggest that these creatures were able to articulate differentiated vocalizations. On the other hand, the fossil skulls also give some indication as to the formation of the speech apparatus. The skulls of the descendants of **Homo erectus** (= upright man), were strikingly similar in this area to those of present day young children (Fig. 56). It is possible that the speech abilities of **Homo erectus** remained at the level of young children for a long time. This suggests, according to the biogenetic rule, that the ontogeny of the human speech apparatus, and also of the speech control centre in the cerebrum, ought to repeat the essential stages of their phylogeny. Human beings are born only with the potential for speech, however, and must learn their native language anew in each generation. The language development of the child appears to repeat the essential phases of early human speech: babbling, articulation and abstraction. The babbling phase is characterized by expressions (crying) of contentment and discontent; the articulation phase by naming of persons and objects with mono- or polysyllabic words, and by construction of sentences. The abstraction phase, finally, is characterized by further conceptual differentiation to still further development, to terminology (--> formal speech -->learned speech).

---

1 The question what makes a human being a human being, e.g. the question of the origin of human existence, cannot be answered by science alone. However, science can investigate the internal and external conditions which accompanied human evolution and influenced its course (Leakey and Lewin 1978).

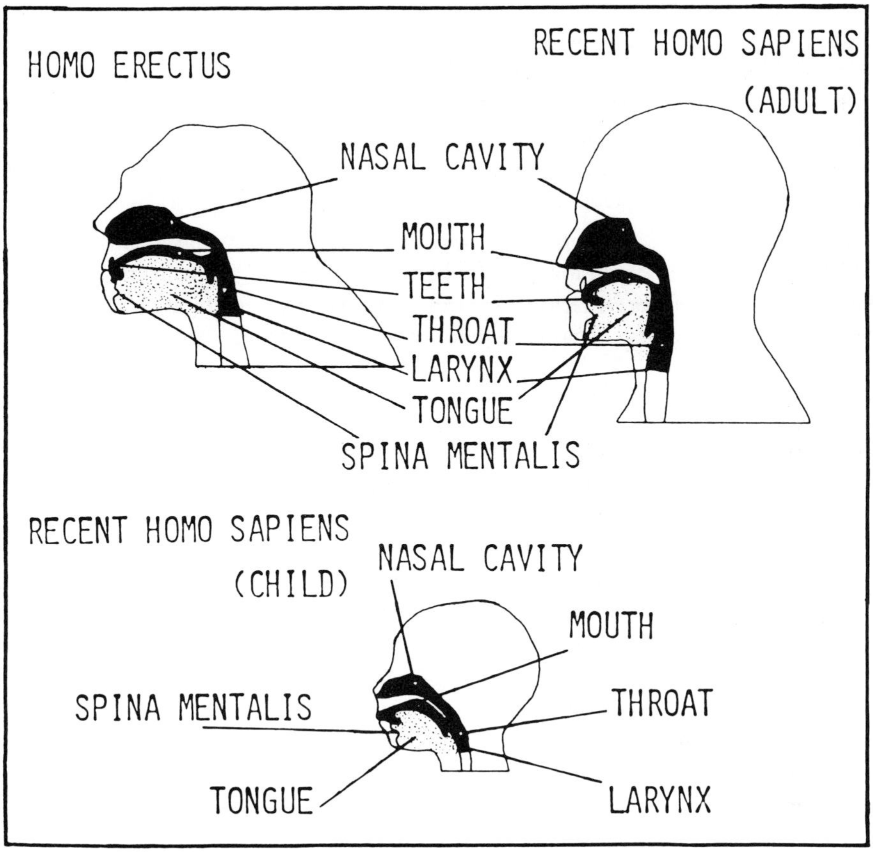

Fig. 56. Comparison of various stages in the **development of the speech apparatus** from primeval to recent human beings (adapted from Hoff and Miram 1979, Liebermann 1984).

IV. Reconstruction

A. The Problem

There are no hints, at least not at our present state of knowledge, indicating that at some early stage of evolution, independent, monofunctional polycytes united in a symbiosis to generate a multifunctional organism. Instead, the data from ontogeny, phylogeny, neurobiology and ethology suggest that the organs of a polycytic organism are possibly of endogenous origin and arose in the course of a long process of differentiation. In this respect, multicellular organisms are presumably elementary organisms and have not evolved through symbiosis of different polycytes.

The biogenetic rule, in principle, describes how the transition from the eucytic unicells to the polycytic organisms and thus to the carriers of culture, to human beings, occurred (see p. 140, 142). This process can be regarded as socialization in the broadest sense, i.e. the formation of associations of cells into organisms, and finally of organisms like human beings into states. Therefore, this entire phase of evolution could be termed "**sociogenesis**".

Cultural evolution, created by humans, appears as a new chapter in the course of biological evolution. **Konrad Lorenz** (1973) coined the term "**fulguration**" for the concept that the leap from animals to human beings could only be explained by special mechanisms of macro-evolution (macro-mechanisms of evolution are discussed in Chapter Six).

B. Fulguration model

1. Polycyte

The first animal or plant organisms were eucyte unicells or protists. As a rule, they reproduced asexually by division of their cells. The daughter cells separated from one another, grew, and divided again. The failure of the cells to separate after division must have led to the existence of a simple polycytic system. When such layers of cells became large enough, they could join to form hollow spheres with one layer of cells. In time, division of labor among these cells must have

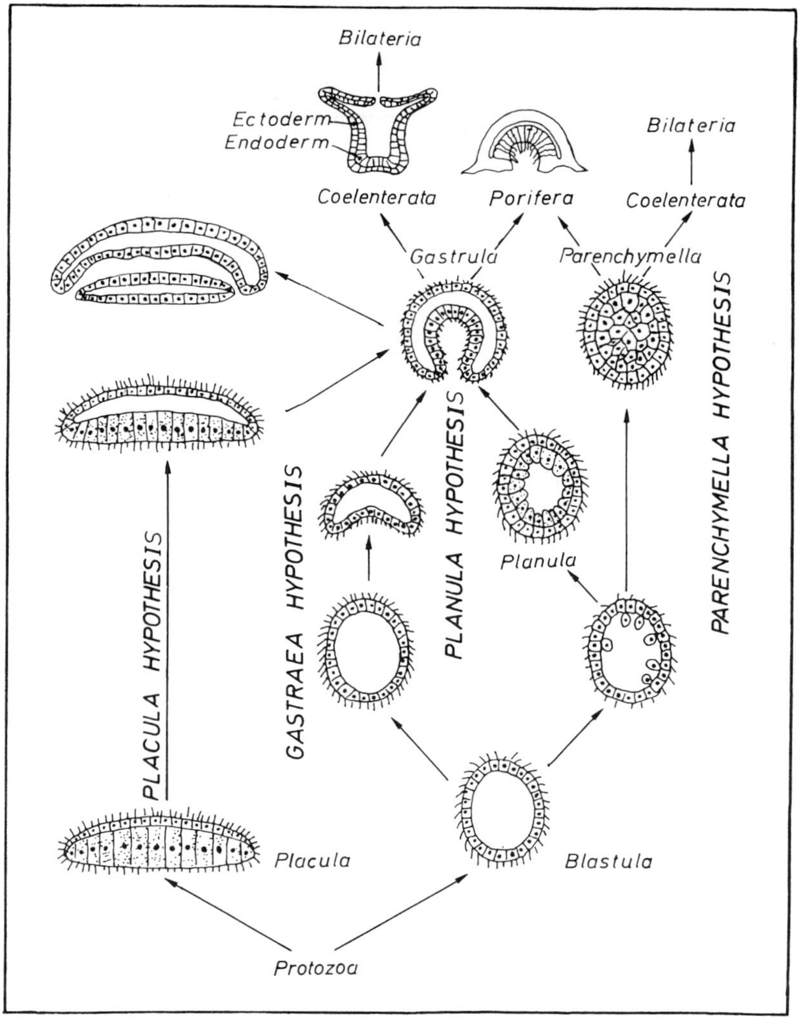

Fig. 57. Hypotheses on the **origin of multicellular animals**. Gastraea hypothesis, E. Haeckel; Planula hypothesis, E. Ray-Lancester; Parenchymella hypothesis, E. Metschnikoff; Placula hypothesis, D. Bütschli (adapted from Kuhn and Probst 1980).

arisen through selection. Hollow spheres of phytoflagellates developed, as **Volvox** has, cells with larger eyespots at one pole (anterior pole). This, together with coordination of the beat of their flagella, permitted the system to move towards the light. At the other end (posterior pole), large, reproductive cells could have differentiated, which preferentially absorbed nutrients. Nutrients could be exchanged with the other cells via cytoplasmic bridges. There are several hypotheses concerning the origin of the first animal polycytes from protozoa (Fig. 57). They all have in common the differentiation first of the ectoderm, and then of the endoderm. Cells deposited between these two layers could finally have led to the formation of a third cell layer; this could have been the raw material for the differentiation of further multicellular systems such as those for transmission of information, delivery of nutrients, removal of metabolic wastes, support and defense. The ever larger variety and number of specialized tissues eventually developed into various organs, which induced the development of a central nervous system for their coordination. In the further course of evolution, the organs were refined and optimized[1]. The brain developed to process the ever-increasing flow of information from the developing sensory organs, and to coordinate the animal's reponses to their more accurately perceived environment.

The human being - due to a highly developed self-consciousness - now managed a further step upward to a new phase of evolution, the cultural evolution.

2. *Human beings*

Like all polycytes, humans are primarily the result of biological evolution. Their phylogenetic development from the higher primates is well documented by fossils. **Ramapithecus** - the name is derived from the site of the first find in India - is regarded as the first ancestor of mankind (Woll 1979). This primate lived in South Asia and East Africa, 14 to 12 million years ago, the climate became drier, and the rainforests receded in favour of savannahs. In sparse forest, it is advantageous for an animal which must cross stretches of ground between trees to be able to run fast and to stand up straight in order to look around. Both prey and predators can be seen earlier from an upright stance. There was thus

---

[1] The reason for the different evolutionary patterns in plants and animals may be that the "open systems" characteristic of more highly developed plants have a multitude of organs of the same type (e.g. leaves), whereas with the "closed systems" of animals, the more highly they are developed, the more limited the number of organs of the same type appears to be (e.g. heart of vertebrates).

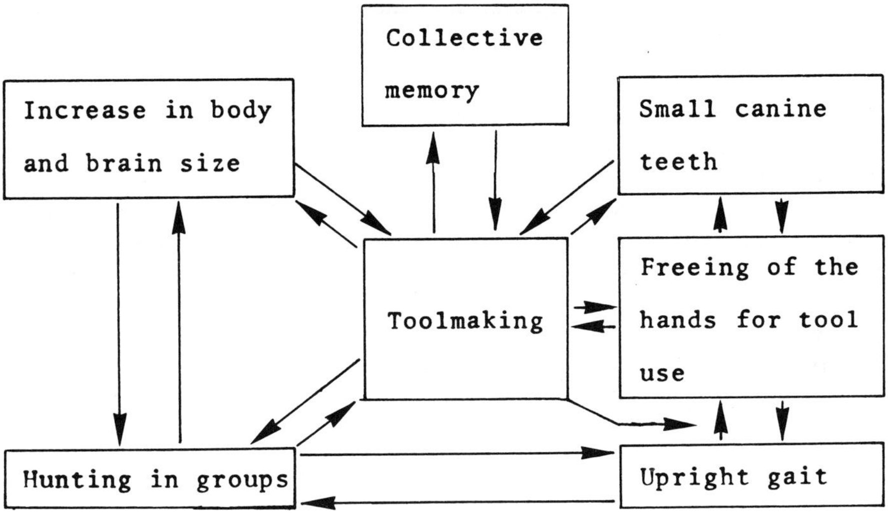

Fig. 58. Correlation between biological and cultural evolution (Leakey and Lewin 1978).

selective pressure toward an upright gait via the known biological micro-mechanisms (compare p. 134). This released the forelimbs for other functions. They could now become hands with which to gather food and to defend against enemies. The relief of the muzzle from its function as the sole organ for processing food and its relief from defense led to changes in the appearance of the face (jaws, teeth, musculature). The head moved into a statically favourable position over the support column, and the brain volume increased up to 400 or 500 ml.

The first unequivocal fossil skulls of **Homo habilis**, with a brain volume between 500 and 800 ml, have been found in East Africa and are 3 million years old. **Homo habilis**, the "tool-using human-like being", used stone tools which were consciously prepared before hunt or defense required them. Such future-oriented behaviour may have been a typical trait of these human ancestors, based on a further development of the cerebrum beyond that of the ur-primates. In addition, the increased capacity of the cerebrum may have led to a limited ability to differentiate among vocalizations (**babbling phase**, Saussure 1967). However, **Homo habilis** seems to have had a relatively small memory capacity and most of his behaviour was probably instinctive, being modified relatively little by learning.

Between 2 and 1.5 million years ago, **Homo habilis** gave way to **Homo erectus**, the fully upright human. **Homo erectus** spread from Africa to Europe and Asia. The appearance of a special bony projection (spina mentalis = "thorn of intelligence"; anchor for part of the tongue musculature) on the inside of the chin of this species indicates a great improvement in its ability to speak (**articulation phase**). At the same time, the brain volume increased from 700 to 1,200 ml, which must have been accompanied by an increase in brain functions, including the formation of a large storage capacity for memory. This would have had an ever increasing influence on behaviour. In addition to the use of tools, **Homo erectus** learned to use fire.

About 1 million years ago, according to fossil discoveries, the immediate ancestors of the recent **Homo sapiens** evolved in Africa. From there, the "reasoning" human expanded over the entire planet. This would have been made possible by a further increase in brain function and thus abilities which accompanied the increase in brain volume to 1,250-1,600 ml, which is the average for **Homo sapiens**. This probably led to further improvements in speech (**abstraction phase**). Tools of stone and bone, funerary gifts, cult items and cave paintings give evidence that the human cerebrum must have developed structures for a consciousness of the environment and of the self in time, and thus paved the way for an independent, creative cultural evolution phase. This is represented by the different ethnic groups living today: e.g. the Aborigines, Mongolids, Negrids and Europids (Caucasians).

3. Culture

Through genetic predisposition and pressure of selection towards a life in hunting bands, the ancestors of mankind were formed into social beings. This promoted the development of a complex communication system of **symbolic language**, which has to be learned individually, rather than being instinctive. Human language means the translation of thoughts into combinations of sounds. Thus language puts the environment into a form which can be mentally manipulated. The history of human language shows a progressive abstraction of concepts from daily life, which eventually led to a formal language, the prerequisite for the development of literature, philosophy, science, law, logic, etc.

The process of language development was made possible by a parallel evolution of the cerebrum, including the differentiation of special speech centres. The human cerebrum also makes it possible for people to become conscious of themselves and of their environment, to remember experiences, and to pass these on to succeeding generations (tradition) independently of genetics. Our long childhood and adolescence enables us to learn not only through trial and error, but also through imitation and from teaching (Woll 1979). Traditions arise which allow each new generation to make use of the experience of earlier generations and to proceed from a higher level of knowledge. Writing, which appeared at least about 4,000 years B.C., is a result of permanent advances in tool use. It makes it possible to store information outside of genes and the brain, and thus to enhance the collective long-term awareness or memory (Fig. 58).

The beginnings of cultural evolution (hunting in groups, toolmaking, language, tradition, writing, etc.) are thus interrelated with biological evolution (upright gait, increase in brain volume, changes in the facial structure, etc.) in complex ways. For example, according to the biological evolutionary micro-mechanisms, the upright gait freed the hands for use and production of tools; in addition, the hands gained an important function for handling food, which affected the development of teeth and jaws (Fig. 58). These differences in motor coordination of forelimbs and upright gait in turn influenced the further evolution of the brain and - in connection with speech and writing - especially the development of the cerebrum. The production of special and ever more precise tools gave mankind the possibility to acquire "**constructed organs**" at need - in contrast to the naturally selected organs of animals - and for this reason the human hand remains unspecialized.

In comparison with biological evolution, which works with an enormous expenditure of time and material, cultural evolution occurs with explosive rapidity. The rate of development is multiplied by the fact that the human brain now delegates information transfer to material objects (e.g. computer): the development of writing, followed by printing, has made it possible to transport information over distance and time. The tempo of cultural evolution has been quite different in diverse populations, due to environment and opportunities for communication of the group, so that at present there are groups living at the cultural level of the stone age (e.g. Bushmen as the oldest living culture) simultaneously with some highly technological societies. On the other hand, technologically advanced societies are seen by some as being spiritually and psychologically impoverished, compared to certain "stone-age" cultures.

V. Future development of mankind

The enormous development of the cerebrum led to the evolution of mankind, and through cultural evolution to a new quality of life. **Human beings, possibly the only organisms conscious of themselves, progressively began to adapt the natural world to their own will, rather than adapting to it.** For about 11,000 years, humans have intervened in the biological evolution of the environment by breeding domestic plants and animals (Hoff and Miram 1979). They become ever more persistent in settling areas with unfavourable climates, exterminating animal competitors for food, and combating parasites and disease through the use of medicine. Human beings, in their quest for an ever "better" life style, have attempted to harness nature and protect themselves from its vagaries. This has led to an increase in life expectancies and survival rate in children. The result is a gigantic population explosion.

With the realization of atomic fission, mankind has penetrated deep into the material workings of the world. Through medicine and the recently developed genetic technologies, man is even beginning to control his own biological evolution and that of other organisms. Mankind has withdrawn itself and other organisms ever more strongly from the natural mechanisms of evolution. It seems as if biological evolution has reached its end in modern mankind, which is now occupied with a systematic withdrawal from the substrate of its existence. Humanity is now in the deepest crisis since its origin. The decisive difference between biological and cultural evolution implies that the latter is not biologically

---

Fig. 59. Analogy between biological and language evolution (Kull 1986).

ORGANISMS	LANGUAGES
Evolutionary change by many slight, almost imperceptible mutations.	Evolutionary change by many slight, almost imperceptible variations.
Occasionally extensive mutations, phenotypically recognizable.	Occasionally extensive changes, e.g. sound shifts, within a short time.
Adaptation to the environment by establishing in a niche.	Adaption to the environment by establishing in a lingual niche (formation of concepts depending on the environment).
Adaption is never completed.	Incompleteness of every language and its rules.
Partial isolation is followed by emergence of geographical races.	Partial isolation is followed by emergence of dialects and sociolects.
Isolation is followed by emergence of new species, by way of different developments in seperate biotopes.	Isolation is followed by emergence of new languages, by way of different developments in seperate biotopes. (e.g. Scandinavian languages).
Convergence as a consequence of adaption to the same needs on the bases of differing basic forms.	Convergence of languages of different derivation: evolves in groups with similar cultural organization (e.g. all hunting peoples have a great number of concepts dealing with hunt); results also from formation of technical terminology (e.g. medicine).
Gene flux by migration of individuals into other populations.	Language flux by integration of words from another language (foreign words and loan-words).
Mixture in the border area of two races.	Mixed dialects in the border area of two dialects.
As a rule no fecundity between two different species.	In the border area of two different languages there is bilingualism and no mixed language as a rule.
Living fossils.	Extinct languages alive (e.g. language of the bushman people; antiquated German on the linguistic islands of Trentino and Pennsylvania).

heritable and therefore can get lost very easily. Hitler's National Socialism may serve as an example of a "culture" that preserved its inheritance of technology but not its intellectuality or spirituality (e.g. Christianity, Enlightenment, classical period). The generations today must therefore decide whether **Homo sapiens** may live like his ancestors for further millions of years, or risk extermination from this planet as a result of a "mistake of nature". Our present duty seems to be a critical appraisal of the situation and a response to it by developing alternatives. We must ask anew about the origin and aim of our development. We must come to an understanding with technology, find out whether we should do everything we are technically capable of doing, or whether we are bringing about our own destruction.

VI. Micro-mechanisms of cultural evolution

Although cultural evolution of human beings is analogous in some ways to biological evolution (Fig. 59), it differs in essential points, according to **Osche** (1972):

> *1. There is no inheritance of acquired traits in biological evolution; new favourable mutations and gene combinations spread slowly through the population, as a result of selection, because they can only be transmitted to the offspring of the individuals which possess them. Cultural evolution consists entirely of transmission of acquired traits. Through language, and later writing and modern communication devices, discoveries of an individual can rapidly become the common possession of large parts of humanity. Cultural evolution, therefore, occurs much more rapidly, and this is the reason why today we have, simultaneously, aboriginal peoples who live at the level of the stone age, and a highly developed technology which makes space travel possible.*
>
> *2. In biological evolution there is adaptation of traits to the conditions of the environment. In cultural evolution, human beings adapt the environment to their will (needs).*

3. In biological evolution, differential utilization of the environment is achieved by adaptive radiation[1] *and formation of different species, each with its own ecological niche. In cultural evolution, adaptation to a niche (e.g. in "professions") occurs through differentiation of the "tools", without speciation. In spite of the numerous ecological niches which human beings are able to create, there is only one species,* Homo sapiens.

4. Biological evolution works opportunistically, and can only make use of success. It cannot "learn" from "failure". The same unfavourable mutations and gene combinations can occur over and over; and in the course of evolution species and groups have become specialized in ways which led to extinction when the environment changed (e.g. dinosaurs and Neandertal man). Cultural evolution also leads to specialization, but human beings as biological organisms remain "unspecialized": they can learn from their mistakes and avoid these in the future.

VII. References

Altmann, S.A. (1967): *Social Communication among Primates*. University of Chicago Press London.

Andrews, P. (1982): *Hominoid Evolution*. Nature 295 185-186.

Childe, V.G. (1975): *Soziale Evolution*. Suhrkamp Wissenschaft, Vol. 115, Frankfurt.

Denffer, D. von, Ziegler, H., Ehrendorfer, R. and Bresinsky, A. (1983): *Lehrbuch der Botanik für Hochschulen*. 32nd edition; Gustav Fischer Verlag Stuttgart, New York

Eder, K. (1980): *Die Entstehung staatlich organisierter Gesellschaften*. Suhrkamp Wissenschaft Frankfurt.

Goleman, D. and Davidson, R.J. (1979): *Consciousness. The Brain, States of Awareness and Alternative Realities*. Irvington Publ. New York.

Griffin, D.R. (1982): *Animal Mind - Human Mind*. Dahlem Konferenzen. Springer Verlag Berlin, Heidelberg, New York.

Haeckel, E. (1868): *Natürliche Schöpfungsgeschichte*. Gemeinverständliche wiss. Vorträge über die Entwicklungslehre. Jena.

Hoff, P. and Miram, W. (1979): *Evolution*. Schroedel Verlag Hannover.

---

[1] Definition p. 178

Kahle, W. (1986): *Nervous system and sensory organs*. Thieme Verlag Stuttgart.

Kuhn, K. and Probst, W. (1980): *Biologisches Grundpraktikum*. Vol. II; Gustav Fischer Verlag Stuttgart, New York

Kull, U. (1979): *Evolution des Menschen. Biologische, soziale und kulturelle Evolution*. Volume 6, J.B. Mettler, Stuttgart.

Langman, J. (1977): *Medizinische Embryologie*. 5th revised edition; Thieme Verlag Stuttgart.

Leakey, R.E. and Lewin, R. (1978): *Wie der Mensch zum Mensch wurde*. Hoffmann and Campe Hamburg.

Liebermann, P. (1984): *The Biology and Evolution of Language*. Harvard University Press Cambridge, Massachusetts.

Lorenz, K. (1973): *Die Rückseite des Spiegels*. Piper Verlag München.

Montagu, J., Ashley, F. (1962): *Culture and the Evolution of Man*. Oxford University Press New York.

Osche, G. (1972): *Evolution. Grundlagen -- Erkenntnisse, Entwicklungen der Abstammungslehre*. Studio visuell, Herder Verlag Freiburg.

Popper, K.C. and Eccles, J.C. (1977): *The Self and its Brain (An Argument for Interactionism)*. Springer Verlag Berlin, New York, London.

Portmann, A. (1969): *Biologische Fragmente zu einer Lehre vom Menschen*. Schwabe Verlag Basel, Stuttgart, 3rd edition.

Remane, A. (1952): *Die Grundlagen des natürlichen Systems, der vergleichenden Anatomie und der Phylogenetik*. Akademische Verlagsgesellschaft Geest und Portig Leipzig.

Riedl, R. (1980): *Biologie der Erkenntnis. Die stammesgeschichtlichen Grundlagen der Vernunft*. 2nd edition; Paul Parey Verlag Hamburg, Berlin.

Romer, A.S. (1986): *The vertebrate body*. W.B.Saunders Company, Philadelphia, 6th edition.

Saussure, F. de (1967): *Grundlagen der allgemeinen Sprachwissenschaft*. Walter de Gruyter Verlag Berlin, New York

Schilcher, F.v. and Tennant, N. (1984): *Philosophy, Evolution and Human Nature*. Routledge & Kegan Paul London.

Schwartz, J.H. (1984): *Hominoid Evolution - a Review and a Reassessment*. Current Anthropology 25 655-672.

Searle, J. (1984): *Minds, Brains and Science*. The 1984 Reith Lectures, BBC London.

Sengbusch, P. von (1979): *Molekular- und Zellbiologie*. Springer Verlag Berlin, Heidelberg, N.Y. 553.

Woll, E. (1979): *Evolution*. CVK Biologie Kolleg Bielefeld.

Wolpoff, M.H. (1982): *Ramapithecus and Hominid-Origins*. Current Anthropology 23 501-522.

Chapter 6 Macro-Mechanism of Evolution

In transitions from one basic type to another, for example, from procyte to eucyte, the evolutionary step is larger than speciation, and supposedly occurs through special **macro-mechanisms** of evolution (e.g. Kuhn 1976, Stanley 1981, Illies 1983). As long as quantitative results regarding the macro-mechanisms of evolution are unavailable, the word "principles" should rather be used in this context, the more so, as the principles known until now are generally only valid with certain reservations. The principles of macro-evolution will be presented in the following.

I. Phase principle

The evolution of simple systems to more complex ones is not a continuous process. Rather, it is characterized by a constant oscillation between **divergent** and **convergent phases**. The phase principle derived by **Kuhn** 1976 from chemogenesis was later expanded to apply to the entire process of evolution (Schwemmler 1979). In the divergent phases, the systems of a given evolutionary level become more varied and fill the pre-existing "posts" (niches) without increasing in complexity.

Divergent phases are followed by convergent phases which are characterized by increases in the degree of complexity of their evolutionary systems. They represent the qualitative "leaps" in evolution. The most important convergent and divergent phases of the entire evolutionary process of cosmo-, chemo-, bio- and sociogenesis are summarized in Fig. 60.

II. Modular principle

The modular principle (Schwemmler 1972, 1984) describes the way by which a qualitative "leap" from one basic type to another could have occurred in the course of evolution. **This states that each next higher evolutionary stage is achieved by integration of variants of the next lower level**, each of which is very well adapted to a different ecological niche. The result is a hierarchy of interlocking evolutionary levels, and each successive level is of a higher degree of complexity. The levels are hypothetically summarized in the scheme of Fig. 61 (see also Fig. 60). Thus, atoms arose from various particles; the primitive gases from various atoms; the biologically important molecules from various primitive gases; macromolecules from molecules; mixed aggregates (prebio-organelles) from various macromolecules; a minimum organism, the precyte, from various prebio-organelles. In the further course of evolution, procytes arose from various precytes and, according to the Endocytobiological Cell Theory, eucytes arose from integration of different procytes; and finally the polycytes such as plants, fungi and animals, including human beings, arose from different eucytes. Mankind then became the source of the cultural phase of evolution. The modular principle is thus in a general sense a symbiotic mechanism (**symbiogenesis**). In addition to mutation and selection, symbiogenesis is an important mechanism of evolution (**Schwemmler** 1972).

Whether evolutionary transitions from one basic type to another occur, following an inner compulsion (Bresch 1977) or as a result of chance (Monod 1970), is a long standing scientific controversy. With the insight that natural laws control chance, **Eigen** and **Winkler** (1976) have shown a way to resolve the controversy. Although random variation and selection produce individual structures, their combination to form basic types appears to be the result of a causal evolutionary process. Neither of the two opposing views can be proven clearly at present. However, an underlying periodicity in the property of the smallest components of complex systems hints at causality behind evolutionary events. This principle has been introduced as the periodicity principle (Schwemmler 1984).

Fig. 60. Representation of the hypothetical phase principle of the overall evolutionary process, which involves constant oscillation between divergent and convergent phases. The rectangles represent the convergent phases, the open spaces in between the divergent ones.

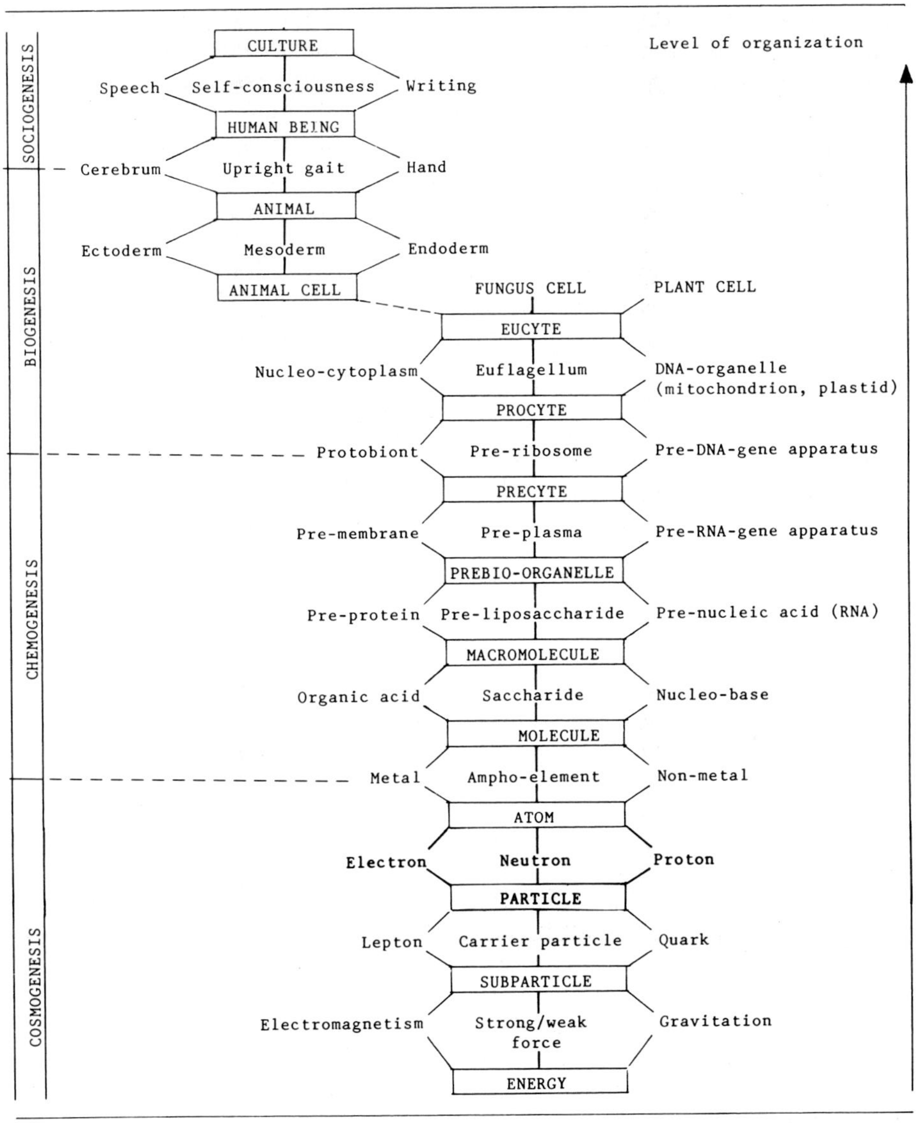

Level of organization
SOCIOGENESIS
BIOGENESIS
CHEMOGENESIS
COSMOGENESIS
CULTURE
Speech
Self-consciousness
Writing
HUMAN BEING
Cerebrum
Upright gait
Hand
ANIMAL
Ectoderm
Mesoderm
Endoderm
ANIMAL CELL
FUNGUS CELL
PLANT CELL
EUCYTE
Nucleo-cytoplasm
Euflagellum
DNA-organelle (mitochondrion, plastid)
PROCYTE
Protobiont
Pre-ribosome
Pre-DNA-gene apparatus
PRECYTE
Pre-membrane
Pre-plasma
Pre-RNA-gene apparatus
PREBIO-ORGANELLE
Pre-protein
Pre-liposaccharide
Pre-nucleic acid (RNA)
MACROMOLECULE
Organic acid
Saccharide
Nucleo-base
MOLECULE
Metal
Ampho-element
Non-metal
ATOM
Electron
Neutron
Proton
PARTICLE
Lepton
Carrier particle
Quark
SUBPARTICLE
Electromagnetism
Strong/weak force
Gravitation
ENERGY

III. Periodicity principle

The **periodicity** of material systems has been thoroughly studied in the case of atoms. The various atoms can be arranged in a **periodic system of the elements** developed by **Meyer** and **Mendeleyev** in 1869. Since the periodic system is well known, only the general principle of arrangement will briefly be presented in the following. It is based on the quantum theory (A. **Einstein**, M. **Planck**) and the atomic theory derived from it (J. **Schrödinger**, W. **Pauli**), which also explains the origin of the elements by atomic fusion or the counteracting process of nuclear decay (compare. Fig. 5, p. 18). According to this theory, both light and heavy elements arose in the course of physical evolution by fusion of hydrogen nuclei (protons) and electrons or neutrons (compare p. 16-17). The atomic nucleus consists of protons, positively charged, and neutrons; it is surrounded by negatively charged electrons in various energy states (orbitals). The number of electrons or protons is characteristic for each element and increases in the periodic system from left to right, and from top to bottom. There are numerous regular relationships among the elements which are revealed by this arrangement. The chemical properties of the elements vary in a regular pattern as one regards the horizontal rows (periods), the vertical columns (groups) and diagonal relationships. The pattern can be roughly divided into three regions. If a diagonal is drawn from the upper left to the lower right corner of the chart, the metals or electropositive elements, which give up electrons relatively easy, are located below and to the left of the diagonal. The non-metals, which readily take up electrons, are located above and to the right of the line. Along the diagonal there are so-called amphoteric elements which have characteristics intermediate between those of metals and non-metals (Fig. 62).

Various systems have also been established by theoretical physicists (Quigg 1985) for the smallest components of physical evolution, the elementary particles. The properties by which the particles are classified are the rest of the mass, the electric charge and the spin. The system of **leptons** includes the relatively light particles with spin 1/2 (but both right- and left-handed spins are possible), e.g. the electron and its antiparticle. Leptons can be interconverted by force-mediating particles, the **bosons**. They are distinct from the system of heavy particles called hadrons, which include protons and neutrons. **Hadrons** are not elementary particles in the actual sense of the word, however, for they consist of still smaller subparticles, the **quarks** and **antiquarks**, which have opposite partial charges. Like leptons, all quarks have the same quantum mechanical spin of 1/2 (but right- and left-handed) and cannot exist in isolation. Quarks can also be interconverted by force-mediating particles, the **gluons**. Although leptons respond to the electro-weak force, while quarks are subject to the electro-strong

Cosmo-genesis	Particles	+ Particles	--> Atoms
	Atoms	+ Atoms	--> Molecules
Chemo-genesis	Molecules	+ Molecules	--> Macromolecules
	Macromolecules	+ Macromolecules	--> Prebio-organelles
Bio-genesis	Prebio-organelles	+ Prebio-organelles	--> Precytes
	Precytes	+ Precytes	--> Procytes
	Procytes	+ Procytes	--> Eucytes
Socio-genesis	Eucytes	+ Eucytes	--> Polycytes (eg humans)
	Human beings	+ Human beings	--> Cultures

Fig. 61. General representation of the modular principle of overall evolution; the representation does not correspond, neither formally nor with respect to content, to the symbols of chemical formulae.

force, the two groups have so far not been included in a single comprehensive system. In the grand **unified theories**, however, such interconversion between leptons and quarks, by means of special force particles (X and Y **bosons**), has been postulated (Davies 1984). If this should be confirmed by experiments with the next generation of particle accelerators, the physicists are optimistic that they will soon be able to construct a unified system of elementary particles which will include the carrier particle of electromagnetism, the **photon**, and the postulated carrier particle of gravitation, the **graviton**. It is possible that such a system will have a periodic pattern comparable to the periodic system of elements, and that its underlying structure will consist of three basic groups, as is shown in hypothetical form in Fig. 62.

A. A preliminary periodic system of cells

There has been no lack of attempts to arrange the oppressive multiplicity of biological elements, or cells, in a systematic fashion. One of these attempts was made by **Harald Riedl** (1972). In classifying plants, he indirectly sketched a cellular arrangement system as well. However, he distinguishes between cellular classification systems which are based on evolutionary types, and those based on functional types which arise anew at each stage of evolution. He thus describes the process of evolution as a spiral which passes through the same positions (**functional types**) at each higher level (**evolutionary types**). Seen thus, the organizational or evolutionary types of cells correspond to the periods of atoms, and the cellular functional types correspond to the groups of atoms in the periodic system of the elements. So far, Riedl has not gone beyond this general model to a specific categorization of the individual organizational or functional types.

Another attempt to arrange the wealth of cellular forms in a comprehensible order was made by "**Max**" **Taylor** (1974) and **Lynn Margulis** (1976). According to their suggestion, procytes (monera) are called **monads**, because they contain only a single genetic system with a single protein synthesis apparatus (Fig. 63). The full Endocytobiological Cell Theory indicates that mitochondria-lacking unicells without symbionts (ur-protista) are **dyads**, because in addition to their nucleus-cytoplasm system, they are presumed to have a second genetic system from the hypothetical flagellar symbiont. Unicellular animal and fungus systems are **triads**, as they have the additional mitochondrial genetic system . Multicellular animals (metazoa) and fungi could correspondingly be called **polytriads**. Plant unicells are **tetrads**, because of the integration of the plastid protein synthesis apparatus, and multicellular plants (metaphyta), finally, are **polytetrads**.

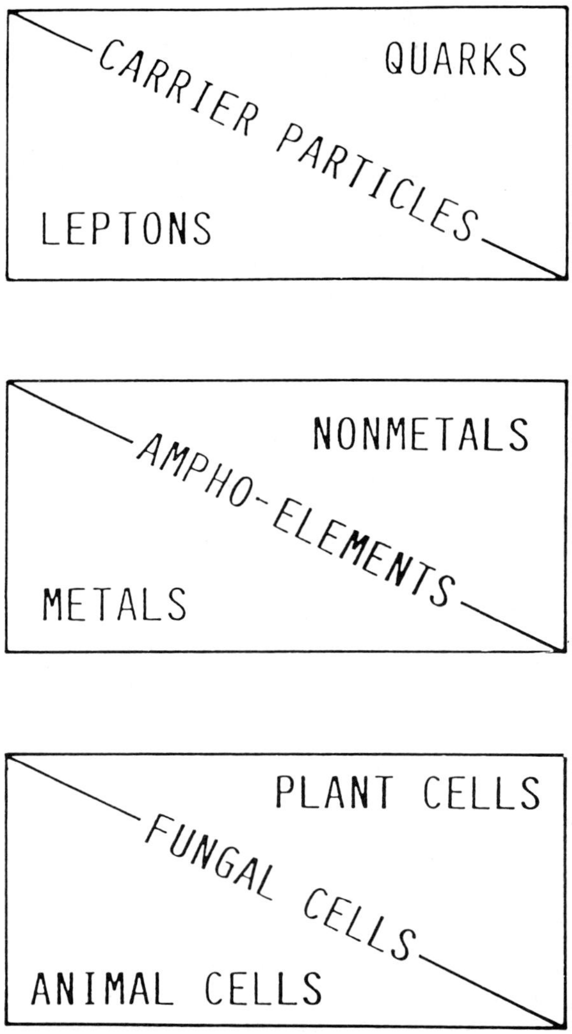

Fig. 62. Representation of the hypothetical periodicity principle of evolution. The phases of evolution appear to build three-fold, periodic patterns of their smallest components.

Depending on the integration of further genetic systems or protein synthesis apparatus by formation of endocytobioses (symbiogenesis), pentads, hexads, heptads, etc. are formed.

On the basis of the complete Endocytobiotic Cell Theory, the framework of Riedl and the terminology of Taylor and Margulis can be used to construct a **periodic system of cells** (Schwemmler 1984), to be regarded as a preliminary system. A final periodic system of cells can be laid down, following systematic comparisons of DNA and RNA nucleotide sequences of each cell type as well as comparisons of sequences of its endocytobionts or genetically semiautonomous organelles.

In the following, a periodic system of cells will be presented for discussion. This system has four horizontal periods in the cellular "main groups". The periods represent different, but related, i.e. homologous, evolutionary types. The procyte monads (bacteria, bluegreen bacteria) make up the first period, the "symbiont-free" flagellated unicells without mitochondria (hypothetical group or ur-protista), as dyads, make up the second period. Trigenomic animal and fungus cells make up the third period, and tetragenomic plant cells, the fourth (Fig. 64). The animal/fungus cells and the plant cells are each divided into two subperiods; one contains the monotriadic or monotetradic unicells (protista), and the other contains the multicellular polytriadic and polytetradic animals and plants.

The comparison of the cell types in the periods reveals eight vertical groups of cellular functional types which are analogous in their special requirements for environment and substrate. In general, we distinguish fermenting, respiring, photergic and photoassimilating functional types of cells (compare p. 72, 78). The biotope for the fermenting cell type can be described as dark, organotrophic, chemotrophic and C-heterotrophic; for the respiring cell type, as dark, organo-/inorganotrophic, chemotrophic and C-heterotrophic or C-autotrophic; for the photergic cell type, as light, organotrophic, phototrophic and C-heterotrophic; and finally, for the photoassimilating type, as light, organo-inorganotrophic (anaerobic) or inorganotrophic (aerobic), phototrophic and C-autotrophic. Each of these functional types can in turn be subdivided into anaerobic and aerobic sections, which thus produce the eight vertical groups of the preliminary periodic system of cells. In the first and second group we find the fermenting cells, some of which contain proplastids; these have growth and absorption functions. Examples of these are gametes, embryonic, blastema, intestine and root cells, and the zooflagellates and zoociliates. Groups three and four include respiring cells capable of differentiation, some of which have support functions and some of which have etioplasts. Examples of these are spore, muscle, blood, bone, cartilage, phloem, wood and storage cells, and theamoebas and sporozoans.

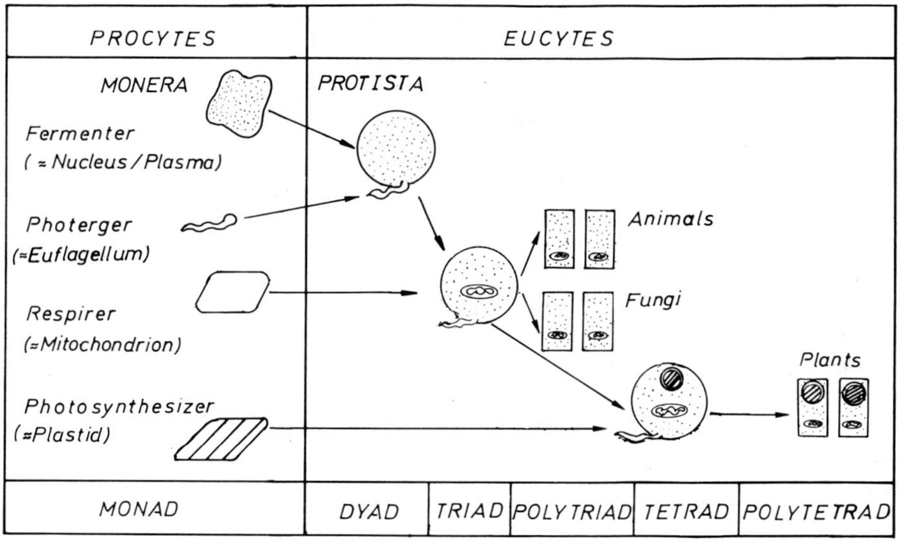

Fig. 63. Classification of procytes and eucytes according to the Endocytobiotic Cell Theory with respect to the increasing number of genetic systems or protein-biosynthesis apparatus (adapted from Taylor 1974 and Margulis 1976).

These groups also contain chemoautotrophic cells whose biotope is dark, organo-/inorganotrophic, chemotrophic and C-autotrophic. In groups five and six are the photergic, pigmented, light-responsive cells, such as the rod-cells of the retina, chromoplast-containing blossom cells and halobacteria. The seventh and eighth group, finally, include all photosynthesizing cells, some of them containing chloroplasts: lichens, leaf cells, photobacteria and bluegreen bacteria. Uptake and integration of additional fermenting, respiring, photergic and photoassimilating endocytobionts into animal, fungus and plant cells give rise to the pentadic, hexadic and heptadic periods of the cellular "subgroups" which will not be discussed here in greater detail (details are given by Schwemmler 1984). The cellular subgroups show that the evolution of the eucyte is far from complete. It is proceeding constantly as new endocytobionts (cf. cyanelles) are taken up into the nucleus-cytoplasm system of the eucyte and, after a long period of coevolution, are integrated as a kind of additional DNA-containing organelle (compare p. 132, 134). One would not, however, conclude that an increasing number of DNA-containing protein synthesizing apparatus of the eucyte increases its degree of high development. A eucyte containing too many organelles of symbiotic origin could be at an evolutionary disadvantage.

As can be seen in Fig. 64, the periodic system of cells - like that of the elements - displays not only many **horizontal** and **vertical** relationships, but also characteristic **diagonal relationships**. If a diagonal is laid from the upper left to the lower right, one finds the predominantly photoactive cells, the typical plant cell types above the line, and below it, the predominantly aerobic glycolytic cells, the typical animal cells. In the region of the diagonal are "ambivalent" fungus cells or fungus-like cell types. The latter for example have cellulose walls like plant cells but contain no plastids and are sometimes bounded by chitin, like some animal cells. At least some fungi appear to have evolved from plant cells by loss of plastids, while others may have evolved from corresponding animal cells.

It is evident that not only atoms, but also cells can be arranged in a rationally derived periodic system. According to this system, cell evolution appears as a spiral of constantly increasing numbers of genetic or protein synthesizing apparatus. It is possible, with the aid of this system, to arrange the great wealth

Fig. 64. Preliminary periodic system of cells: main groups (Schwemmler 1984).

Functional types (analogy series)	Groups / Periods	Evolutionary types (homology series): I, II — FERMENTATION Anaerobic Aerobic Dark Organotrophic Chemotrophic C-heterotrophic Absorption-growth type	III, IV — RESPIRATION Anaerobic Aerobic Dark (In)organotrophic Chemotrophic C-hetero(auto)trophic Sclero-differentiation type	V, VI — PHOTERGISM Anaerobic Aerobic Light Organotrophic Phototrophic C-heterotrophic Light-mobility type	VII, VIII — PHOTOSYNTHESIS Anaerobic Aerobic Light (In)organotrophic Phototrophic C-autotrophic Photoassimilation type
1	MONADS PROCYTES Cytoplasm	Precytes Fermenters (e.g. fermenting bacteria)	Respirers (chemoautotrophs, e.g. sulfur respirers)	Photergers (e.g. halobacteria)	Photosynthesizers (e.g. photobacteria, bluegreen bacteria)
2	DYADS Mono- UR-EUCYTES dyads Cytoplasm Poly- Euflagellum dyads	Mixotricha ? (without symbionts)	Archamoeba? (without symbionts)	?	?
3	TRIADS ANIMALS Mono- FUNGI triads Cytoplasm Euflagellum Poly- Mitochondria triads	Trichoprotozoa Absorption, growth cells (intestine, embryonic, blastema cells; gametes)	Sporoprotozoa Myxomycophyta Support, differentiation cells (cartilage, bone, blood, muscle cells)	Pigmented cells, fungal fruiting body cells (rod cells: eye, nerve cells)	Lichens (algae and fungi)
4	TETRADS Mono- PLANTS tetrads Cytoplasm Euflagellum Poly- Mitochondria tetrads Plastids	Cells with proplastids (embryo, root cells)	Unicells lacking plastids (e.g. fungal cells) Cells with etioplasts, amyloplasts (stem, phloem, wood, storage cells)	Chromoflagellates (yellow, brown algae) Cells with chromoplasts (blossom cells)	Phytoflagellates (green algae) Cells with chloroplasts (e.g. leaf cells)

of cellular forms in homologous evolutionary or analogous functional types; in addition, the system gives a functional-causal understanding of the complex interactions between the nucleus-cytoplasm system on the one hand, and of the DNA-containing cell organelles or endocytobionts on the other. In addition to that, on the basis of the cellular periodic system, it is possible to predict the nature and existence of endocytobiotic cell systems which are not yet known, and to produce them artificially (e.g. photoassimilating animal cells produced by infection of fibroblasts with chloroplasts). However, it must be emphasized that this preliminary periodic system of cells is independent of the eventual refutation or confirmation of the Endocytobiological Cell Theory since the construction of such a system on the basis of the classic cell hypothesis would look quite similar. The various physiological niches and protein biosynthesis apparatus which are the criteria for classification would be the same, regardless of their exogenous or endogenous origin. However, this preliminary periodic system seems more logical and more convincing on the basis of the Endocytobiotic Cell Theory.

B. Possible periodicity in cultural genesis

Periodicity appears to be typical of chemical and biological evolution. It is postulated, therefore, that the physical evolutionary process preceding chemogenesis, and sociogenesis, following biogenesis, would not behave differently. Keeping this in mind, periodicity should be regarded as a general phenomenon of evolution, characterized by repetitive radiation and adaptation at the highest level. The cause for this may be that the basic conditions remain the same over the entire process of evolution, and thus it may lead to the occupation of the same "posts" with ever higher evolutionary levels. Obviously this leads repeatedly to analogous periodic patterns. Therefore, we may ask whether the periodicity principle applies to **cultural evolution**. We can basically distinguish the Oriental cultural groups with a Mongolid cast, including past and present American Indians; the eastern and western Occidental cultural groups originating from Europid cast; those African cultural groups of a Negrid cast; and last, though not least, the individual cultural groups of the Aboriginal races.

If the periodicity principle should be confirmed by future studies for all four evolutionary phases, all the periodic patterns would show a typical subdivision into three parts (cf. Fig. 62), which elsewhere has been called the triality principle (details see Schwemmler 1984).

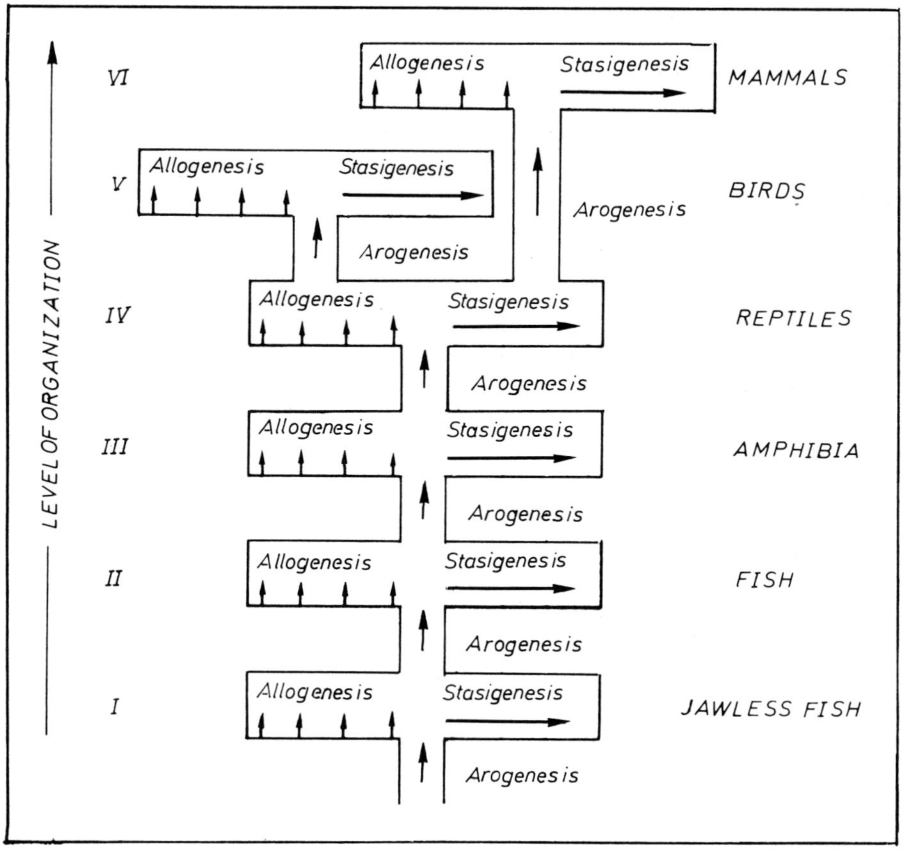

Fig. 65. Pattern of micro- and macro-evolution in the vertebrate phylogenetic tree. Each box corresponds to an adaptive phase (micro-evolution). The transition from one adaptive phase to the next is characterized by evolutionary novelties (macro-evolution: evolutionary "leaps"): I-->II: jaws; II-->III: lungs, legs; III-->IV: egg shells; IV-->V: thermal homeostasis, feathers; IV-->VI: thermal homeostasis, secondary jaw joints, hair, milk glands (adapted from Kuhn and Probst 1980, Ax 1984).

IV. Consequences for biogenesis

The micro-mechanisms discussed at the end of the chapter on biogenesis, explain sufficiently the origin of new species or races, mainly by mutation and selection. However, questions remain about the interaction of various evolutionary factors, and these are often still unexplained. The term micro-evolution was coined for relatively small evolutionary steps, which are not characterized by increases in complexity. Biological micro-evolution is characterized by a great divergence of a basic type into numerous variants, caused by the process of selecting actual ecological "posts". This phenomenon is called **adaptive radiation** (cf. the divergent phases of the total evolutionary process, compare p. 165). Other names are allogenesis, allomorphosis, cladogenesis and idioadaptation. All of these terms describe the same process, namely the differential adaptation of contemporary representatives of one organizational group to various ecological niches. This process can occur by speciation within a geologically short period of time, without the complexity of organization tending toward increase. A well known example of adaptive radiation is provided by the Darwin finches of the Galapagos Islands. There are a number of species which specialize in different types of food; all appear to have a common, seed-eating ancestral form which probably migrated from the South American mainland at a geologically recent time. Such adaptive radiations are often well documented by intermediate forms. Proceeding lines of development are well testified, an example is the evolutionary series of fossils in the development of horses from five-toed to one-toed animals. Such an evolution series is called gradualism.

A less frequently occurring, second form of gradual evolution would be the long-term developmental lines which have diverged little, and which do not result in a new adaptive radiation. These processes are also a form of biological micro-evolution and lead to the so-called "living fossil lines" (stasigenesis).

As a third form of evolutionary process, finally, there are rare cases in which genetic changes in a small group of species or in one species "lead" it into a new adaptive zone (Kuhn and Probst 1980; cf. the convergent phases of the evolutionary process). Such transitions usually occur relatively rapidly. As soon as one group has made this "leap", a new adaptive radiation begins . This rare form

---

Fig. 66. Representation of molecular and cellular evolution according to the modular principle (Schwemmler 1984; for details see text).

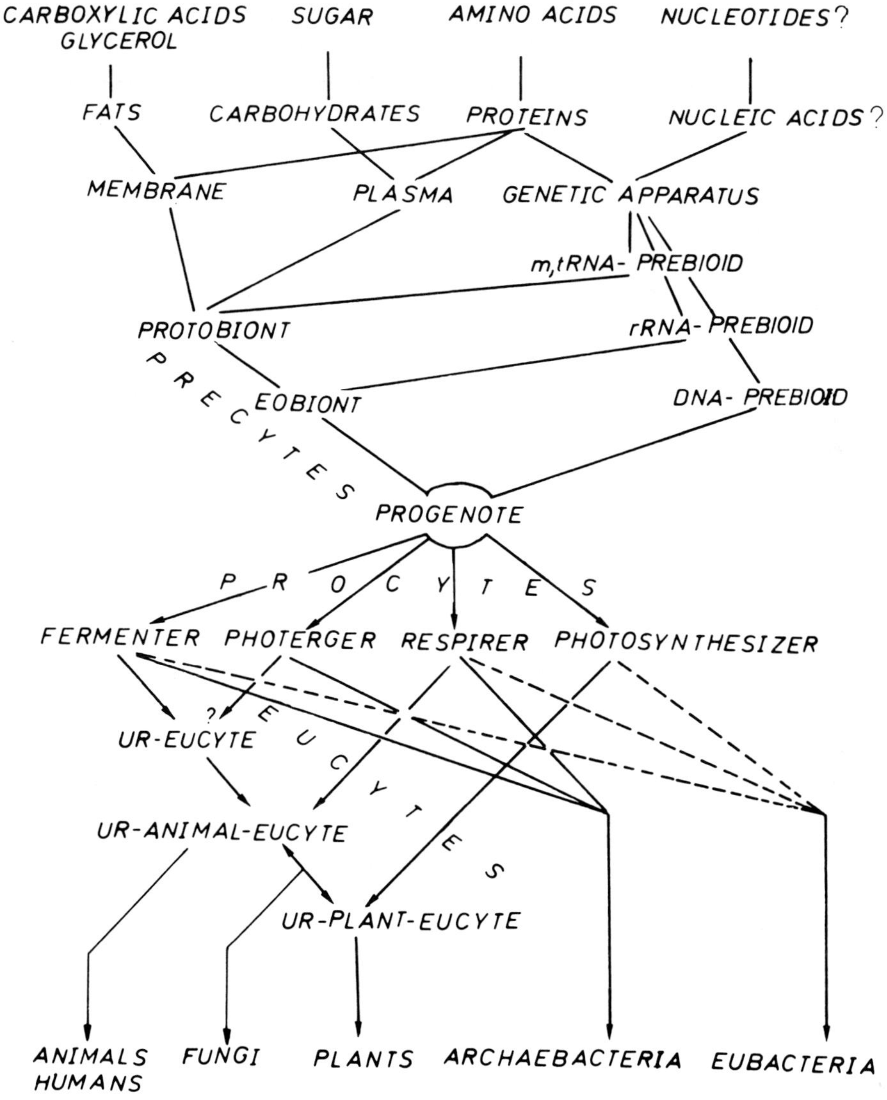
CARBOXYLIC ACIDS
GLYCEROL
SUGAR
AMINO ACIDS
NUCLEOTIDES?
FATS
CARBOHYDRATES
PROTEINS
NUCLEIC ACIDS?
MEMBRANE
PLASMA
GENETIC APPARATUS
m,tRNA- PREBIOID
PROTOBIONT
rRNA- PREBIOID
EOBIONT
DNA- PREBIOID
PRECYTES
PROGENOTE
PROCYTES
FERMENTER
PHOTERGER
RESPIRER
PHOTOSYNTHESIZER
?
UR-EUCYTE
EUCYTES
UR-ANIMAL-EUCYTE
UR-PLANT-EUCYTE
ANIMALS
HUMANS
FUNGI
PLANTS
ARCHAEBACTERIA
EUBACTERIA

of evolution has been named "arogenesis" by **Rensch**. The phenomenon of relatively short-term, evolutionary "leaps" is called punctualism (Stanley 1981). An example is the transition from procyte to eucyte. The term "biological macro-evolution" has been coined for such transitions from one basic type to another; the special principles of macro-evolution discussed above come into play here. In Fig. 65 the various forms of biological micro- and macro-evolution are demonstrated for a simplified, hypothetical phylogenetic tree of vertebrate evolution.

V. Summary

The probability that life on earth arose from non-living but "preconditioned" matter borders on certainty. The ability of matter to become organized into living systems is due to its own inherent capacity for self-organization. The process was, and still is, driven and controlled by the various principles of micro- and macro-evolution. The presence of particular principles of macro-evolution can be verified, at least for portions of the total process of evolution, in the form of the modular, phase and periodicity principles. Yet it is improbable that these principles would not be applicable to the other phases of evolution. On the contrary, the thesis may be put forward that the modular, the phase and the periodicity principles are to be regarded as universal phenomena and to be taken as a basis for any attempt to reconstruct macro-evolution from atom to mankind (Fig. 66). The total process of evolution from big bang to man can be divided into four phases: physical, chemical, biological and cultural (Fig. 67). The essential milestones which had been passed in the course of this process were the atom, the molecule, the cell and the human being. The biological phase of evolution is thus only an intermediate stage in a long, causal chain of events leading eventually to mankind (compare p. 159).

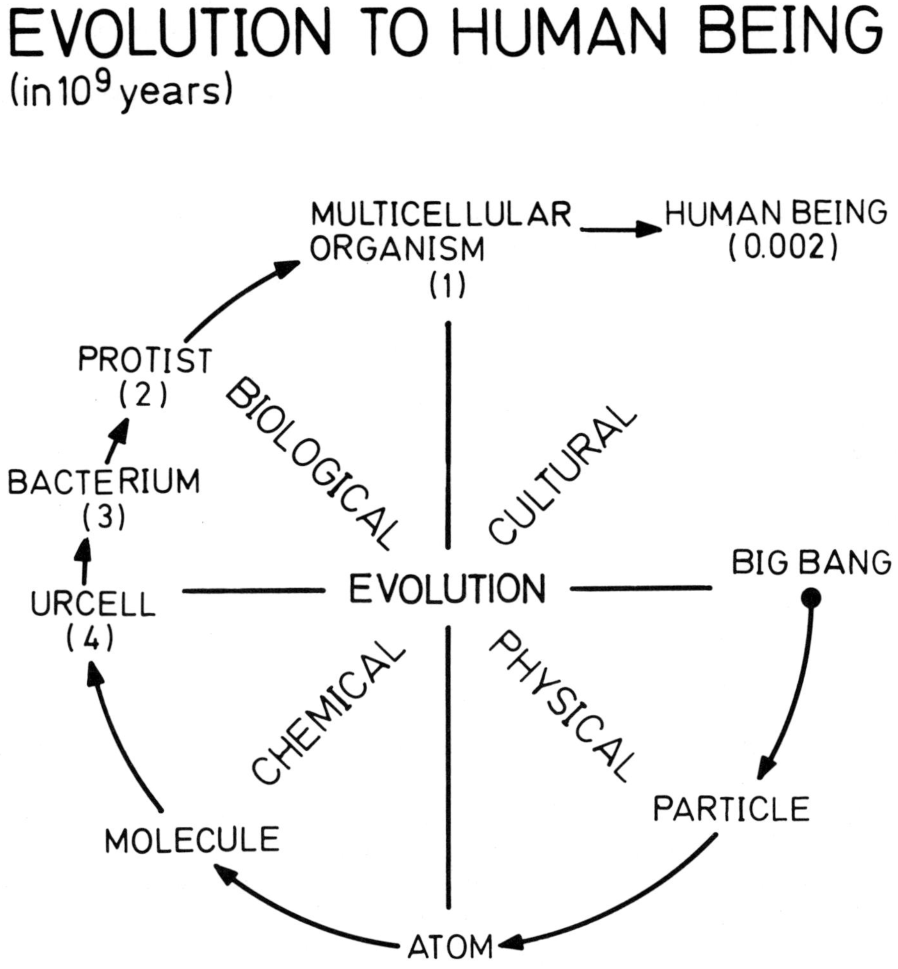

Fig. 67. Hypothetical overall picture of evolution from big bang to man, depicted as an evolutionary spiral (Schwemmler 1984). On the basis of the pulsation theory it can be supposed that after a collapse of matter in the entire cosmos the process of the cosmic, chemical, biological and cultural phases of evolution may recommence, initiated by another big bang.

VI. References

Ax, P. (1984): *Das phylogenetische System*. Gustav Fischer Verlag Stuttgart, New York.

Bresch, C. (1977): *Zwischenstufe Leben: Evolution ohne Ziel?* Piper Verlag München, Zürich.

Davies, P. (1984): *Superforce. Search for a grand unified theory of nature*. Heinemann, London.

Eigen, M. and Winkler, R. (1976) *Das Spiel. Naturgesetze steuern den Zufall*. 2nd edition. Piper Verlag München, Zürich.

Illies, J. (1983): *Der Jahrhundert-Irrtum. Würdigung und Kritik des Darwinismus*. Umschau Verlag Frankfurt.

Kuhn, H. (1976): *Model consideration for the origin of life*. Naturwissenschaften 63 68-80.

Kuhn, K. and Probst, W. (1980): *Biologisches Grundpraktikum*. Volume II. Gustav Fischer Verlag Stuttgart, New York.

Margulis, L. (1976): *Genetic and evolutionary consequences of symbiosis (a review)*. Exp. Parasitol. 39 277-349.

Mayr, E. (1982): *Speciation and macroevolution*. Evolution 36 1119-1132.

Monod, J. (1970): *Le hazard et la nécessité*. Editions du Seuil, Paris.

Quigg, C. (1985): *Elementary particles and forces*. Scientific American 4 64-75.

Riedl, H. (1972): *A model proposed for the progress of evolution with special reference to plants*. Acta Biotheoretica 21 63-85.

Schwemmler, W. (1972): *Endosymbiosebildung: Mechanismus der Evolution*. Naturw. Rdschau 25 (9) 350.

\- (1984): *Reconstruction of cell evolution: a periodic system*. CRC Press Boca Raton, Florida.

Stanley, S. M. (1981): *New evolutionary timetable. Fossils, genes and the origin of species*. New York Basic Books, New York.

Taylor, F.J.R. (1974): *Implications and extensions of the serial endosymbiosis theory of the origin of eukaryotes*. Taxon 23 229-258.

Chapter 7 Experimental Application

I. Endocytobiosis as an intracellular ecosystem

An ecosystem is generally understood to be a well defined area (**biotope**) containing different kinds of organisms living in interdependent communities (**biocenosis**). When the intracellular space represents a biotope in which representatives of two or more different species develop and reproduce in a state of mutual dependence, it may be regarded as an intracellular ecosystem. Like those of the extracellular ecosystem, the individuals of the intracellular ecosystem, per definition, must be genetically independent from each other and self-reproducing. Endocytobioses with autoreduplicative endocytobionts are termed oligogenetic (having several genetically independent systems) intracellular ecosystems (Schwemmler 1980, 1983, Schenk and Schwemmler 1983). The mutual dependence among individuals of an intracellular ecosystem does not necessarily involve an absolute functional specialization of **producers**, **consumers** and **decomposers**; rather, one individual can carry out several functions simultaneously.

By this definition the eucyte cannot be regarded as an intracellular ecosystem. Its DNA-containing organelles are genetically only semi-autonomous, and their reproduction is not self-determined (e.g. the phytochrome-controlled biogenesis of chloroplasts). The eucyte is a multifunctional unit in which the various mechanisms are controlled centrally by the nucleus. **Thus, the eucyte acts as a cybernetic regulatory system. According to the Endocytobiological Cell Theory, the centrally controlled cybernetic system of the eucyte evolved from the oligogenetic intracellular ecosystem of endocytobiosis,** in which the endocytobionts as precursors of euflagella, mitochondria and plastids lost some or all of their genomes to the host cell, possibly by gene transfer.

The phylogenetically younger endocytobioses of leafhoppers, with their semi-autonomous, organelle-like endocytobionts, are a kind of "missing link" between the intracellular ecosystem and the nucleus-controlled regulatory system. The endocytobiotic bacteria of leafhoppers, derived from free-living forms, break down the catabolites (e.g. uric acid) of the host cell (or of the producer) and synthesize anabolites (e.g. amino acids, vitamins) for the nucleus-cytoplasm (e.g. the consumer) and for themselves. These bacterial endocytobionts regulate the osmotic pressure, pH and probably also certain intracellular rhythms of the host cell by gene transfer (Fig. 68).

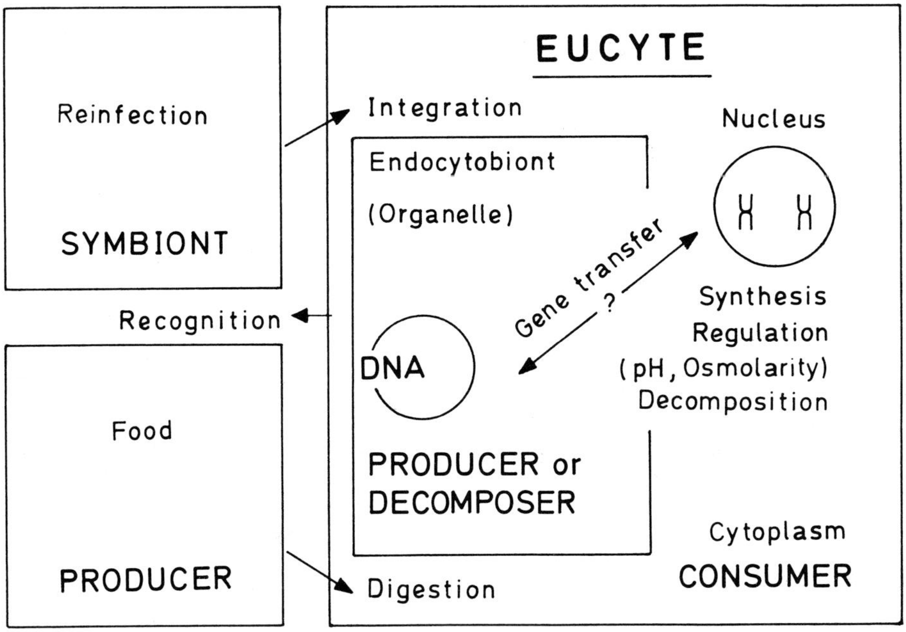

Fig. 68. The eucyte with its semi-autonomous DNA- and RNA-containing organelles as a cybernetic system centrally controlled by the nucleus. According to the Endocytobiological Cell Theory this eukaryotic system evolved from an intracellular, oligogenetic ecosystem. In such systems endocytobionts act, as mitochondria and plastids still do to some degree, as decomposers of the host's nutrients and catabolites (e.g. urea, C-bodies), as producers of anabolites for the host (e.g. vitamins, amino acids and morphogenetic substances), and as regulators of pH, osmotic pressure and probably of cell rhythm of the host cell (Schwemmler 1983).

Leafhopper endocytobiosis is well known as one of the most intensive mutual dependences of host and symbiont, and is thus a good model for analysing the formation of a cybernetic control system at the moment of its inception. This analysis may provide information about the mechanism by which the eucyte cybernetic regulatory system controls a number of phenomena which are still poorly understood: cell differentiation (embryogenesis), **de**-differentiation (tumorigenesis) and circadian rhythm. The egg cell of leafhoppers seems to be especially well suited for such a purpose, because it represents a very simple combined system of the eukaryotic host cell with its prokaryotic endocytobionts which are a structurally separated, symbiotic infectious mass.

II. Endocytobiosis of leafhoppers

The cytoplasm of a mature insect egg is not uniform in structure; a polarity of both morphological and chemical nature is evident (Schwemmler 1980, 1984; Schwemmler and Schenk 1980; Schwemmler and Kemner 1983). Developmental studies suggest that the processes, directly and indirectly connected with the differences between the anterior and the posterior pole, determine at least the beginning of development. The morphology and physiochemistry of the postulated determinants have not yet been elucidated, in spite of intensive research. Solving these problems of cell differentiation will also provide important insight into the problems of cell **de**-differentiation (tumorigenesis) and cell rhythm. Since the polar distribution of determining factors is established during oogenesis, the process of oogenesis will first be examined more closely with the leafhopper **Euscelidius variegatus** as an example.

A. Oogenesis

The ovaries of the female are arranged in pairs, each containing seven ovarioles which open into the vagina. The ovarioles are of the meroistic telotrophic type (Weber 1966). As shown by light- and electron-microscopy, germ cells (oogonia) in the ovarioles form both oocytes and nutrient cells; the latter remain in the nutrient chamber (germarium; Fig. 69). The oocytes move toward the base of the germarium and enter the egg chamber (vitellarium). Here, a cell layer (follicle) forms around them and their cell mass increases enormously. During this period,

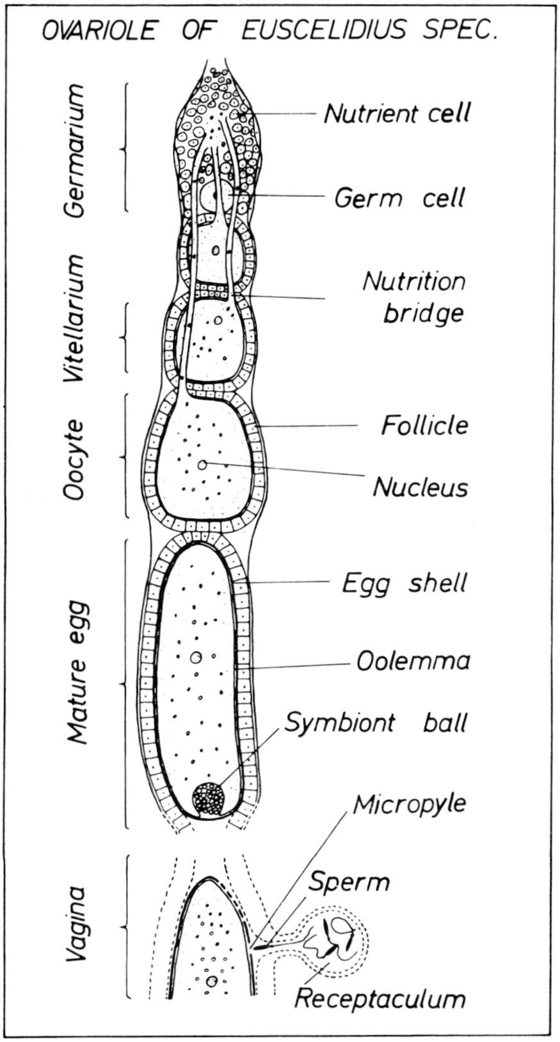

Fig. 69. Formation and fertilization of the egg of the leafhopper **Euscelidius variegatus**. The ovariole is of the meroistic telotrophic type (Schwemmler 1987).

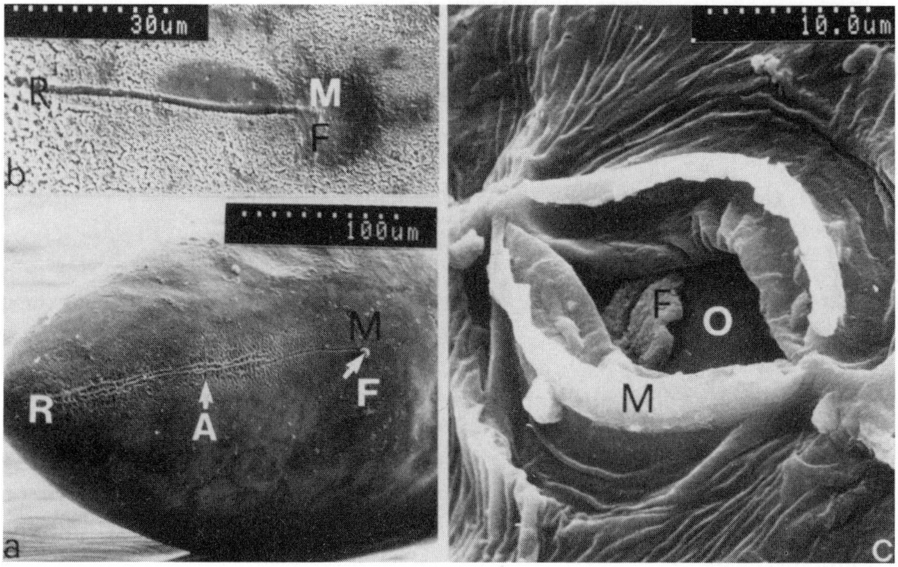

Fig. 70. The entrance opening (micropyle) for the sperm at the anterior pole of the leafhopper's egg, at approximately 75% egg length, as seen under scanning electron microscopy (photograph: U. Gernert, Techn. Univ. Berlin, F.R.G.; Schwemmler 1987).

a. Scored track for hatching of the larva on an unfertilized egg; the micropyle is closed by a lipid plug.
b. Enlargement of a.: micropyle region.
c. Micropyle of a fertilized egg, opened for penetration of sperm.

F = Fat plug, M = Micropyle, O = Egg membrane, R = Scored track.

they remain connected to the germarium by a nutrient strand on their anterior pole. For the further development of the oocyte, yolk globules containing protein, fat, and carbohydrate (glycogen), organelles such as ribosomes, membraneous reticulum (ooplasmic reticulum) and mitochondria move through the nutrient strand into the anterior pole where they are inhomogeneously distributed. An egg develops with a large central yolk section surrounded by a thin, peripheral wall of cytoplasm containing nearly all of the mitochondria. Later the endocytobiotic bacteria attach to the egg at the posterior pole; the egg's external coat forms thereafter and finally the symbiotic infection mass is situated between oocyte and egg coat.

The asymmetrical incorporation, first of the mitochondria at the anterior pole and later of the endocytobionts at the posterior pole, suggests that these structures may be directly or indirectly connected to the developmental determinants we are looking for. Thus, the specific arrangement of these structures during development of the fertilized egg (embryogenesis) will be of particular interest.

B. Embryogenesis

The egg is fertilized during its passage through the vagina. Sperm stored in a sperm pocket (receptaculum seminis) penetrate a special opening (micropyle) in the egg's external coat by means of lipolytic enzymes; the micropyle is closed by a plug of lipid. The micropyle is located at about 75% egg length or EL (the posterior pole is 0% EL). It can be seen distinctly on a scanning electron micrograph (Fig. 70). The first sperm cell to reach the oocyte's nucleus sheds its tail. The middle part of the tail develops into the centriole, which is important during cleavage, and the apical part forms the male pre-nucleus which unites with the female pre-nucleus to form the zygote nucleus (Weber 1966). The zygote nucleus then begins a series of synchronized cleavage cycles, as shown by histological sections and observations in vivo. The first cleavage occurs about 4.5 hours (at 24° C) after fertilization, thereafter the number of nuclei doubles about every hour, without formation of cell membranes (Fig. 71). After the 7th cleavage the nuclei disperse evenly within the egg, and, starting with the 8th cleavage, they then preferentially move to the periphery of the cytoplasm. In the course of the 14th cleavage, occurring about one day after the egg is laid, differentiation of cell membranes leads to formation of a single layer of cells (blastoderm) at the periphery of the egg.

After about 1.5 days a pair of precursors of the germ anlage forms in the blastoderm layer, and on the second day these are drawn together to form the

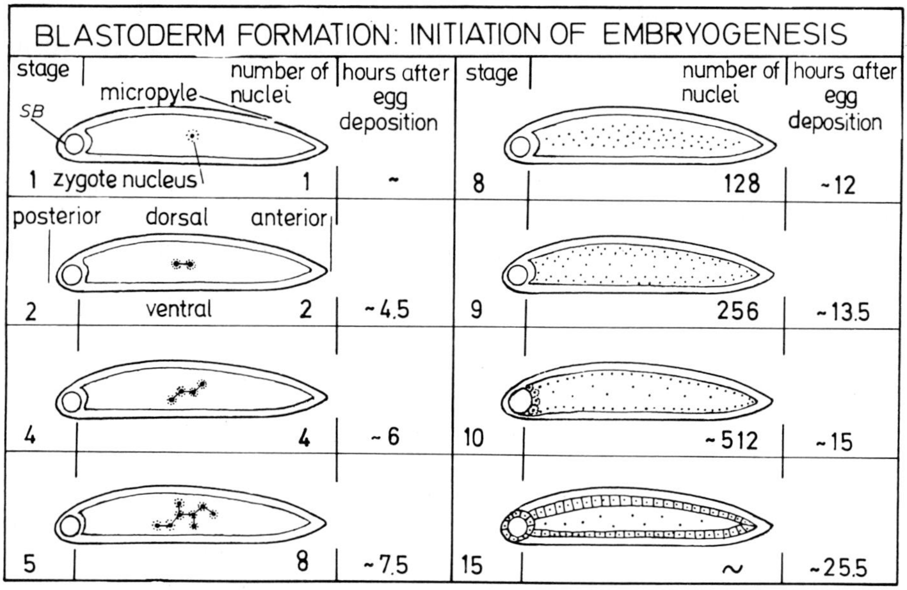

Fig. 71. The first cleavage steps of a fertilized egg up to blastoderm formation. This initial phase of embryogenesis differs clearly from the subsequent phase. The times indicated refer to an incubation temperature of 24^0 C (Schwemmler 1987).

germ anlage (Fig. 72; Zabel and Schwemmler 1980). The germ anlage contacts the endocytobiont mass and then begins to roll forward into the yolk, keeping contact with the mass of endocytobionts. On the third day the germ anlage forms segments, the germ band, in which the basic shape of the body can be recognized. On the fifth day, the germ reaches its greatest expansion and contracts on the sixth day. After the seventh day the embryo stretches out again. On the eighth day, the back closes and the organ anlage differentiates. The larva hatches on the twelfth day, making use of the scored track at the front of the egg (Fig. 70 and 73).

In the course of blastoderm formation, the endocytobiont's mass is first absorbed by the ooplasm and then covered by a cell layer. In the course of germ development, starting on the third day, the endocytobionts are selectively sequestered into special embryonic cells by a complicated process. The descendants of these cells merge later on to form symbiont organs on the right and left of the embryonic abdomen.

The quantitative distribution of mitochondria between anterior and posterior pole show no significant differences up to the time of blastoderm formation. In the anterior pole region, the mitochondria are small and compact, i.e. metabolically active, while those at the posterior pole tend to be long hyaline forms with low metabolic activity (Schwemmler 1987). If infection of the egg is prevented by chemical treatment (lysozyme; Schwemmler and Müller 1986), symbiont-free eggs develop into cephalothorax-embryos without abdomina (Fig. 74; Schwemmler 1973). If, on the other hand, mitochondria are functionally impaired by treatment with antibiotics, some embryos will consist of "headless" malformed abdomina which display tumor-like growth in tissue culture (Schwemmler 1980). This implies that endocytobionts in the leafhopper's egg, directly or indirectly, influence the abdomen-controlling determinants while mitochondria probably have a comparable function for the head. In this way, endocytobionts and mitochondria appear to influence the physiology of the egg, although they operate at opposite ends of the egg and in opposite parts of the embryo. Therefore it is of interest to find out whether the similarity of developmental physiological effects is based on a similar degree of genetic integration of mitochondria and endocytobionts in the oocyte.

C. Gene expression

Analysis of density, sedimentation rates and conformation indicate that the two types of leafhopper endocytobionts contain DNA with a molecular weight between 2.2 and 2.6×10^7 daltons (Schwemmler et al. 1975). If this interpretation

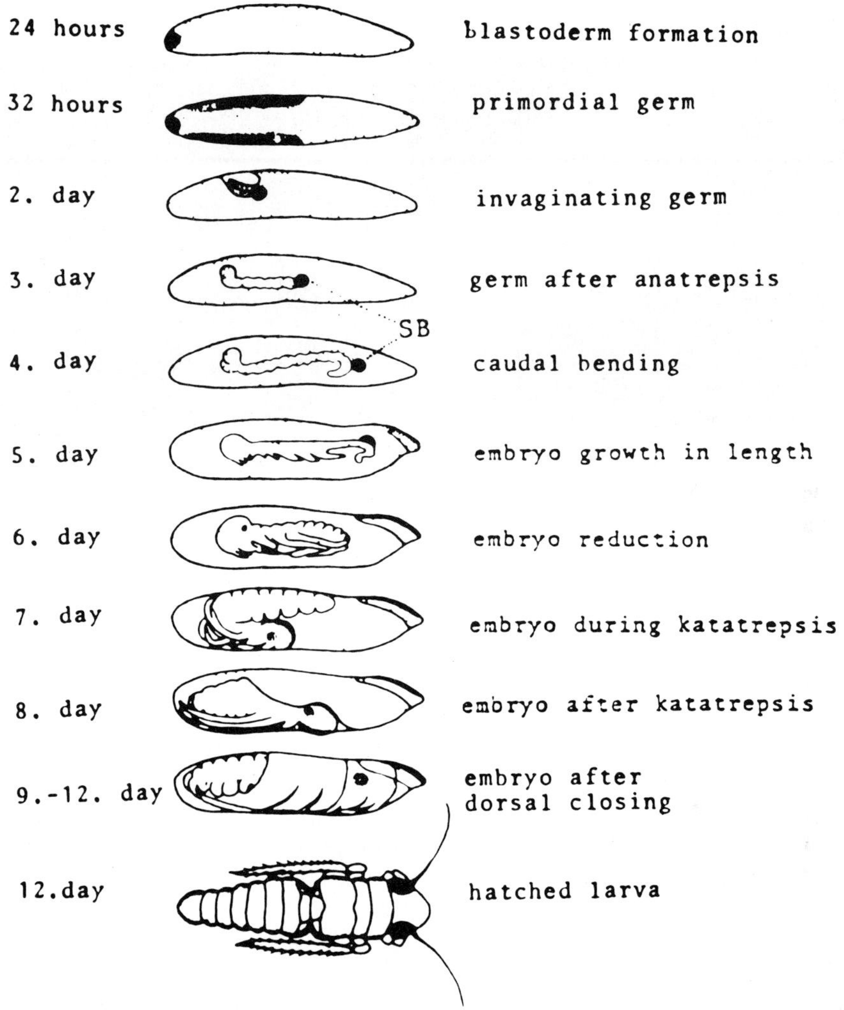

Fig. 72. Embryonic development of the leafhopper after blastoderm formation (Weber 1966, Schwemmler 1987).

SB = symbiont ball.

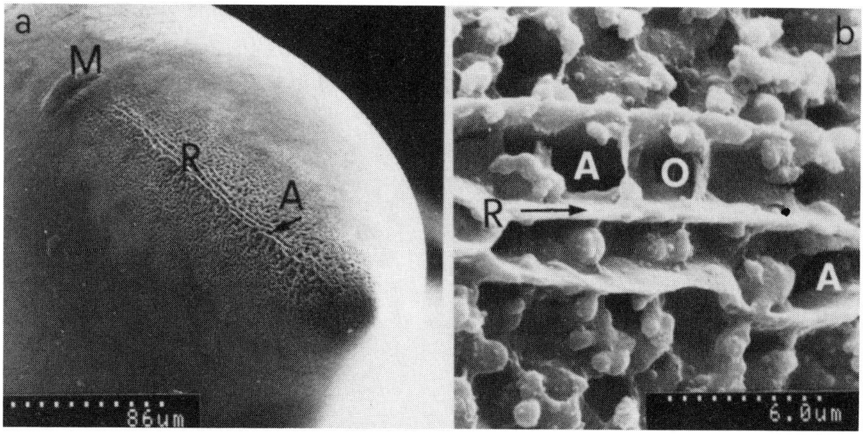

Fig. 73. Irregular double line of egg openings for gas exchange (aeropyles) along the scored track for hatching of the leafhopper larva, located at the anterior pole of the egg (scanning electron micrograph by U. Gernert, Techn. Univ. Berlin, F.R.G.; Schwemmler 1987).

a. Overview.
b. Detail.

A = Aeropyle, M = Micropyle.

of the data is confirmed by experiments now in progress, leafhopper endocytobionts will have the smallest DNA complement of any prokaryote known, only a little larger than that of the mitochondria of leafhopper eggs, which amounts to 10^7 daltons. This low value will be understood if the endocytobiont's genome displays a previously unrecognized degree of overlapping gene expression, if a significant amount of symbiont's genome is transferred into the host cell genome, or if it is lost. Leafhopper endocytobionts, like mitochondria, are incapable of extracellular reproduction in vitro (Schwemmler 1973), which suggests either gene transfer or gene loss. The regulatory coupling between host and symbiont systems, already demonstrated for energy metabolism, pH and osmotic phenomena, may be an indication of gene transfer (Schwemmler 1983). Thus leafhopper endocytobionts seem to be especially suited in studying the intermeshing of gene expression in the host and in the symbiont. Furthermore, they may have been preserved by their hosts for millions of years of coevolution as a kind of living fossil, a "missing link" between mitochondria and archaebacteria. It is likely that gene transfer has occurred between the genomes of symbiont and oocyte, just as it has between mitochondrial DNA and the nucleus, and that this transfer is causally related to the developmental anomalies in the egg which occur when either the endocytobionts or the mitochondria are inactivated. It may now be tested whether the developmental determinants can be discovered by physiochemical experiments.

D. Physiochemistry

At the anterior pole of the leafhopper egg, between about 80 and 100% EL, there is a double track of irregularly spaced pores (aeropyles) which open in the process of hatching and thus serve as a score track. During the development of the egg, they facilitate gas exchange (oxygen and carbon dioxide) with the environment (Fig. 73). Preliminary measurements in microrespiration chambers (Schwemmler 1987) yield the respiration quotient (CO_2 divided by O_2) RQ = 0.92 ± 0.2 per g dry weight of egg (average of 1,000 eggs) for unfertilized eggs, and after the third day an RQ of 0.63 ± 0.2 per g dry weight of egg (average of 1,000 eggs) for fertilized eggs. If the aeropyles and the micropyle (which remains open after the entry of the sperm cells) are artificially sealed, development will stop after the third day. If one assumes that RQ-values greater than 1.5 or twice as large indicate anaerobic glycolysis, values less than 1.5 will indicate aerobic respiration. The average RQ-value of unfertilized eggs of about 1 leads to the assumption that there could be an RQ-value exclusively "oxidative" (degradation of sugar: RQ = 1, or a mixed "oxidative" value; degradation of fat: RQ = 0.6, of

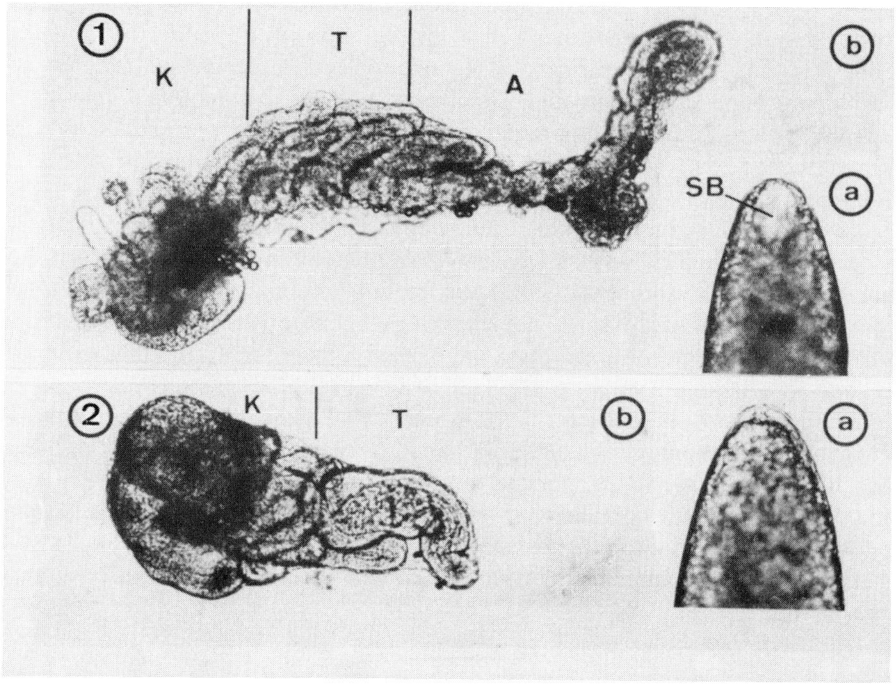

Fig. 74. Impairment of embryonic development of the leafhopper. Embryo approximately 1-week-old (Schwemmler 1973).

1a. Anterior pole of egg with normal symbiont ball.
1b. Normal embryo, developed from the egg of 1a.
2a. Anterior pole of egg without symbiont ball.
2b. Cephalothorax-embryo without abdomen, developed from the egg of 2a.

A = Abdomen anlage, K = Head anlage, SB = Symbiont ball,
T = Thorax anlage.

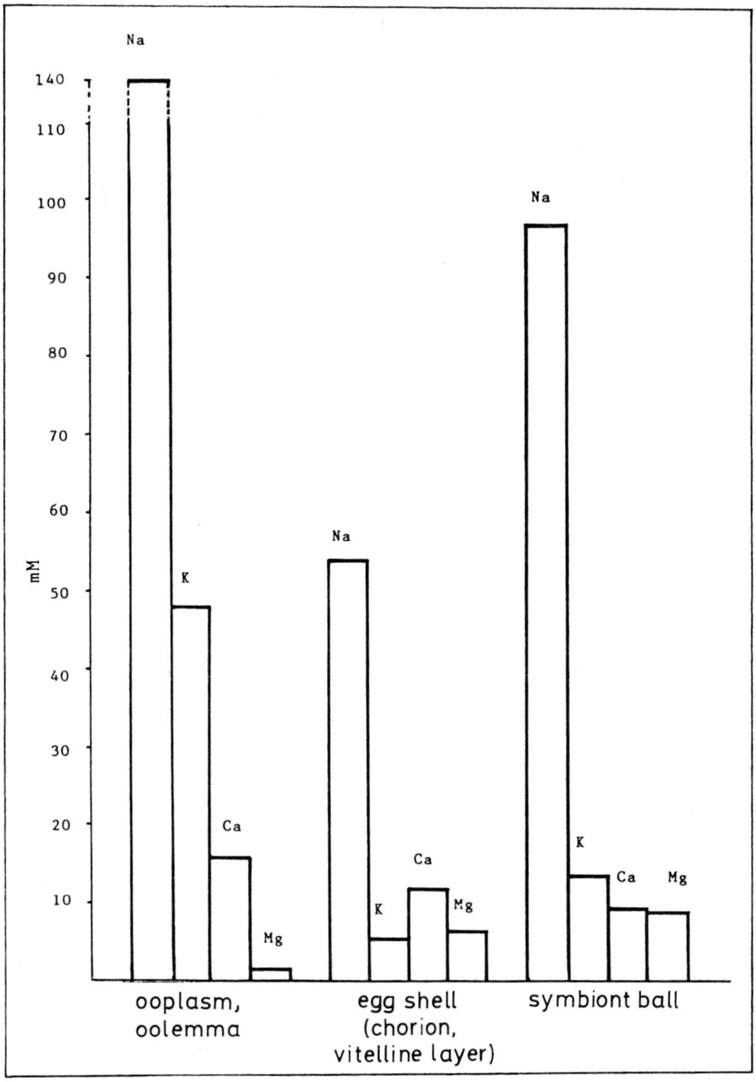

Fig. 75. Cation content of the main egg fractions. Distribution of the most common elements in the various fractions of the unfertilized leafhopper egg, measured by atomic absorption spectroscopy, as an average of three measurement series (three parallel samples each; Schwemmler 1987).

proteins RQ = 0.8, of organic acids RQ = 1.2) or a value derived "oxidatively" and "glycolytically" (degradation of fat 0.6; fermentation of e.g. carbohydrates RQ = 1.5-2). Conclusively glycolysis in the fertilized egg (compare bypass respiration; Kraepelin 1983) is greatly reduced, at first in favour of anaerobic respiration and later in favour of aerobic respiration.

To test this conjecture, various respiratory inhibitors were applied either externally to the egg or injected into its interior. An application of inhibitors of the alternative or bypass respiration (benzohydroxamic acid, $1.5x10^{-3}M$) and of respiratory inhibitors of the cytochrome chain (antimycin A, potassium cyanide and rotenone, each at a concentration of $10^{-3}M$) for three to seven hours did not visibly affect the development of the egg. However, leaving the eggs with the respiratory inhibitors for three days resulted in a complete inhibition of their development after the third day. This seems to confirm the original conjecture that, at the beginning of egg development, glycolytic activity dominates, but that at the third day after oviposition it is strongly reduced in favour of respiratory activity. The question now arises as to where, exactly, within the leafhopper egg then glycolytic and respiratory activities are located which so far have only been indirectly measured.

Preliminary analyses of elements (by atomic absorption and flame emission spectroscopy; Schwemmler 1987) of the ooplasm, egg coat and egg symbionts in unfertilized eggs have yielded the distribution of bound and free elements as follows (Fig. 75):

$$\text{Na } 71\% > \text{K } 18.7\% > \text{Ca } 7.2\% > \text{Mg } 2.9\%$$

For example, an average total concentration of bound and free calcium of about $2x10^{-3}M$ was determined, which corresponds well with values obtained by X-ray microanalysis. X-ray analysis also shows that in fertilized eggs (in contrast to unfertilized eggs) there is a gradient in the total calcium concentration (free and bound), with the highest value at the anterior pole and the lowest value at the posterior pole (Fig.76). Microinjection of fluorescent pH markers (pyranine, fluorescein and neutral red, each at $3x10^{-5}M$) into the abdomina of females shortly before oviposition, or directly into freshly laid eggs, followed by ultraspectrophotometry or densitometry, shows a minimum pH of 6.4 in unfertilized eggs, and a maximum pH of 6.8 in fertilized eggs. Both fertilized and unfertilized eggs have a pH gradient between the anterior and posterior pole (Fig.77). In addition, by use of a bioluminescence method, the anterior pole of unfertilized eggs is found to have a higher ATP content than the posterior pole. Finally, substrate tests, able to detect small amounts of lactate or malate dehydrogenase, revealed activities only at the posterior pole, but succinate

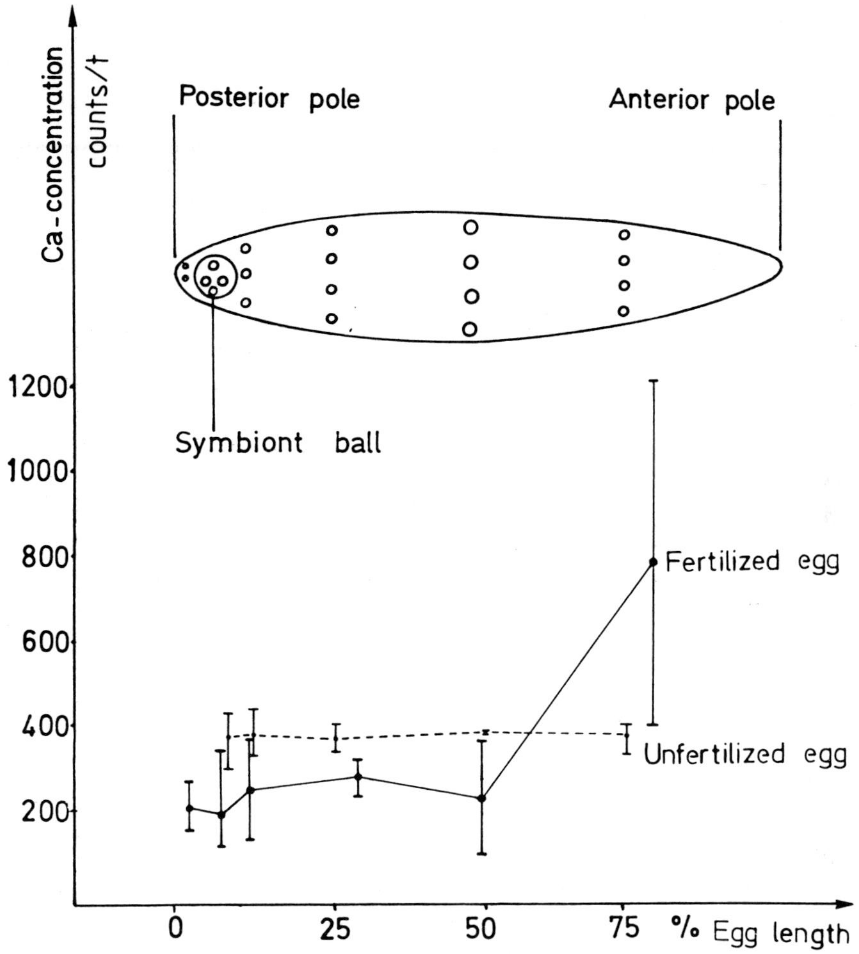

Fig. 76. Relative calcium distribution between the anterior and posterior pole, each as the average of at least four measurements by X-ray microanalysis (Schwemmler 1987).

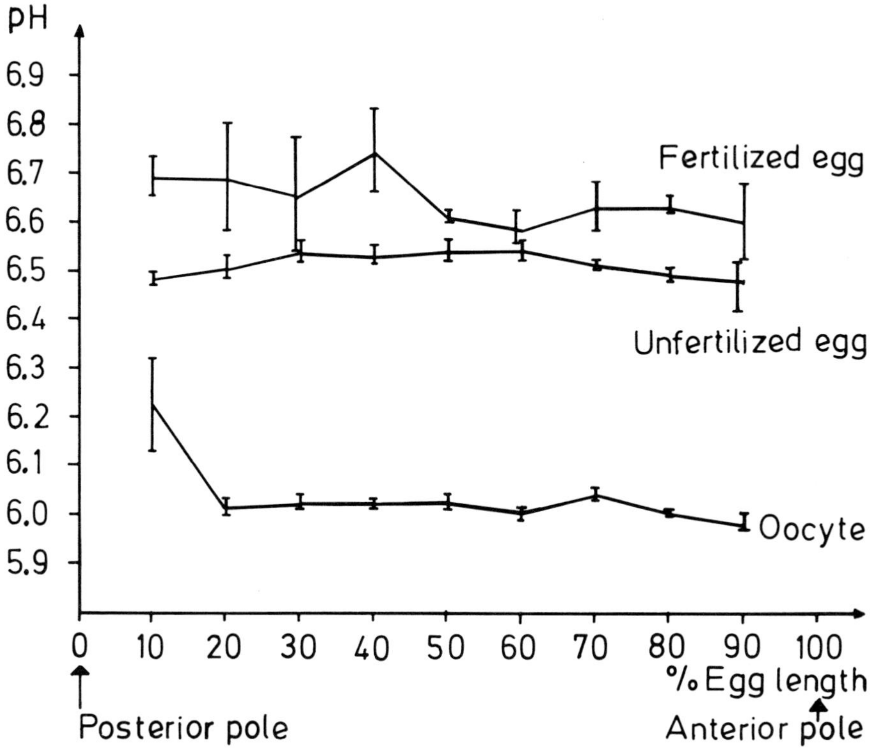

Fig. 77. Topography of pH in oocytes, unfertilized and fertilized leafhopper eggs between the anterior pole (100% egg length) and posterior pole (0% egg length). The graph represents the means (with standard deviation) of fertilized and unfertilized eggs, and of oocytes.

Pyranine was the fluorescence probe, which was injected into the abdomina of female leafhoppers. The fluorescent eggs were photographed at two excitation wavelengths (λ_1 = 410nm, λ_2 = 460nm, λ-emission 490nm) and the negatives were evaluated by densitometry (Agfapan vario XL film, 22-23 DIN). The intracellular variation of measured pH values, and intracellular variation between the three different egg types was less than 0.05 pH units; however, there is an element of uncertainty of about 0.4 pH units in the ordinate values, depending on the calibration curve used. The pH values are reproducible.

dehydrogenase activities at the anterior pole (Schwemmler 1987).From all these physiochemical data we may conclude that in unfertilized eggs, glycolytic activities dominate in the ooplasm, with a maximum at the posterior pole, where the glycolytically active endocytobionts have a direct or indirect effect (Fig. 78). In the fertilized egg, it appears that mitochondrial respiratory activities gradually increase at the expense of glycolytic activities, starting at the anterior pole, and by the third day after oviposition, respiratory activity dominates. Morphologically, the transition from glycolysis to respiration is manifested as a gradual change of the large hyaline mitochondria to small compact forms, starting at the anterior pole.

Thus the determinants deduced from morphological and genetic data correspond to those from physiochemical analyses. We are following the working hypothesis proposed by our research group that the **initiation of leafhopper egg development is controlled by the interaction between ooplasmic glycolysis and mitochondrial respiration** (Schwemmler 1980, 1987; Schwemmler and Schenk 1980; Schwemmler and Kemner 1983). Experiments are in progress to determine whether the hyaline form of mitochondria contains the complete respiratory chain. If it does not (in contrast to the compact form of mitochondria, which does), our working hypothesis will be supported.

E. Model system

There are close structural, functional and genetic analogies between the essential bacterial leafhopper endocytobionts and their mitochondria. Both mitochondria and endocytobionts have decisive functions in the metabolism and in the development of the nucleus-cytoplasm system. These analogies suggest that mitochondria have developed from formerly independent endocytobionts, or that the endocytobionts have acquired the character of DNA-containing cell-organelles. Mitochondria and endocytobionts apparently regulate the interaction between glycolysis and respiration in the host cell. This interaction seems to be part of the basic character of each eukaryotic cell (Schwemmler 1980). It appears to stabilize the cell rhythm, to initiate the development of the egg, and it may also induce tumor-like growth when disturbed (Fig. 79).

An understanding of the molecular interactions of the whole complex may be gained from dynamic analysis of the glycolytic and respiratory components of a single eukaryotic cell. The simultaneous measurement of temporal changes of the glycolytic and respiratory components of a eukaryotic cell in parallel experiments, however, can rarely be accomplished (it is possible with the alga **Acetabularia**), because there are no suitable methods for distinguishing or separating the two

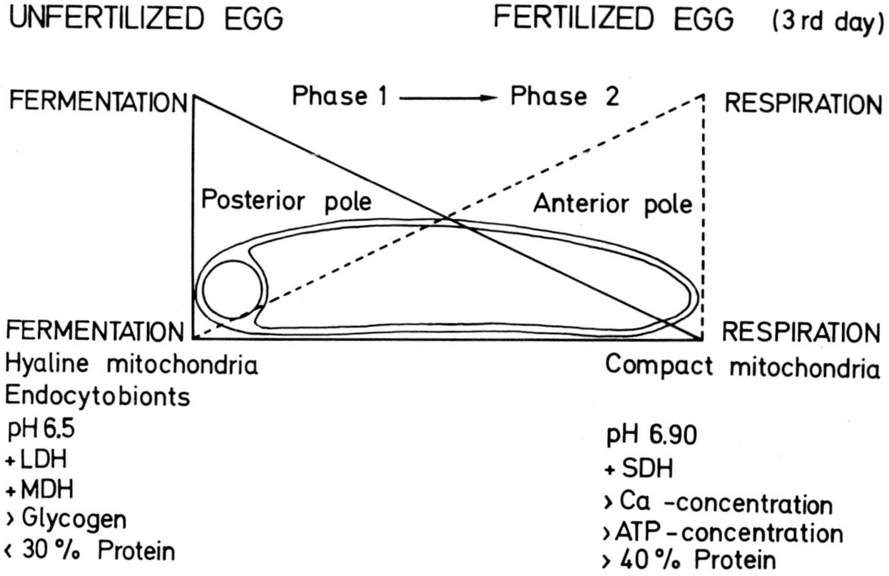

Fig. 78. Working hypothesis for determination of egg polarity.

The working hypothesis presented here is that in the ooplasm and in the endocytobionts of unfertilized leafhopper eggs, glycolysis dominates, with maximum activity at the anterior pole of the egg (phase 1); however, following the 3rd day after oviposition, the glycolytic activity is gradually replaced by respiration of compact mitochondria, starting at the anterior pole (phase 2; Schwemmler 1987).

types of components without damage to the function of the two compartment fractions. On the other hand, the glycolytic and respiratory components of the endocytobiotic system are generally easy to distinguish or separate. Endocytobioses, like those of leafhoppers, can therefore serve as a model for molecular analysis of the eukaryotic cell. Thus endocytobiosis research has much to contribute to cell research, and vice versa. It might even lead to an understanding of the unsolved riddles in cell biology, such as cell differentiation (e.g. embryogenesis), cell **de**-differentiation (including tumorigenesis) and cell rhythm (e.g. circadian rhythm). **The new interdisciplinary research field, endocytobiology, is already being established from a fusion of symbiosis and cell research.**

III. Endocytobiology as an interdisciplinary research field

The term **symbiosis** was introduced to literature by **De Bary** in 1879. It originally included not only **mutualism** (beneficial interdependence of different species), but also **parasitism**, a form of association which is beneficial to one partner and detrimental to the other. At present the term symbiosis is frequently equated with the concept of mutualism. The term is also used in Fig. 80 in this sense. The phylogenetically younger partner is generally considered to be the host, and the older partner the symbiont.

Around the turn of the century, the **symbiont theory** was formulated. According to this, plastids (**Schimper** 1883) and mitochondria (**Altmann** 1890) and later also the nucleus (**Mereschkowsky** 1910) were proposed to be the final products of a long process of symbiotic integration. When DNA was discovered in plastids (Ris 1961) and in mitochondria (**Nass** and **Nass** 1963), symbiosis research received new impetus to achieve more understanding of cell evolution. This is also reflected in the book "Symbiosis in Cell Evolution" by **Lynn Margulis** (1981), foundress of the modern endosymbiont hypothesis. She postulates that in addition to mitochondria and plastids, the eucytic motility apparatus (termed undulipodium by L. Margulis), including the euflagellum, is the result of a symbiotic integrational process (series endosymbiont hypothesis: **Taylor** 1974).

The founder of systematic symbiosis research, **Paul Buchner**, investigated the function of endosymbionts (1965). He assessed their significance in their contribution to the breakdown of nutrients for the host, and in the provision of vitamins to the host which has a limited diet. Later on, this concept had to be developed further for a number of intracellular endosymbionts to account for

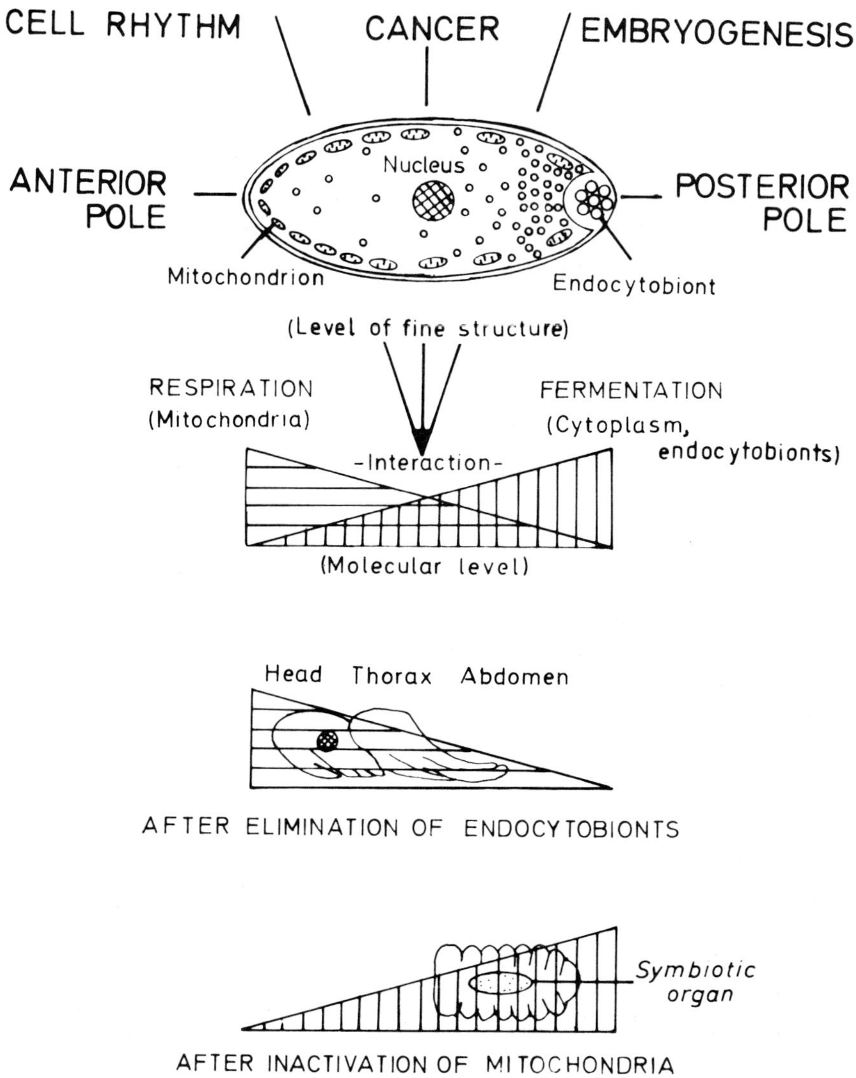

Fig. 79. The leafhopper **Euscelidius variegatus** is a suitable model for research in the fields of endocytobiology.

physiological and biochemical studies which showed that intracellular endosymbionts take on some functions of DNA-containing cell organelles (Fig. 68; Schwemmler 1980b).

There was thus a need for a clear terminology in order to distinguish the various types of symbiosis (Fig. 80). The term **cytobiosis** introduced by **Taylor** (1980), includes not only intracellular but also epi- and intercellular symbiosis, and it is not clear whether it completely excludes extracellular symbiotic relationships. The term **endosymbiosis** frequently used by **Buchner** is equally vague. The term **cytosymbiosis** coined by **Sitte** (Kleinig and Sitte 1984) encounters the same difficulty, especially because it does not include cytoparasitism according to the definition given above. However, the term **endocytobiosis** fits well; **Schnepf** was the first to use it at a conference of the Deutsche Forschungsgemeinschaft (German Society for Research) in 1975. In 1979 the term was elaborated by **Schwemmler** to mean both the intracellular symbionts (=endocytosymbionts: viruses, bacteria, algae etc.) and the intracellular parasites (=endocytoparasites: viruses, bacteria, fungi etc.), thus taking into account the fluid transition between symbiosis and parasitism. In addition the term also includes DNA-containing cell organelles according to the **Endocytobiotic Cell Theory** (an expansion of the series endosymbiont hypothesis). From endocytobiosis Schwemmler derived the term **endocytobiology** (1980b). In 1980 and 1983, at the first and second International Colloquium on Endocytobiology in Tübingen (F.R.G.), Schenk and Schwemmler proposed to officially introduce the term and its definition, in order to establish it internationally at the conference of the N.Y. Academy of Sciences in 1986 (Lee and Fredrick 1987). The term endocytobiology has now become generally accepted for cooperative works of symbiologists, parasitologists and cytologists investigating intracellular symbioses. Furthermore, the cooperation of these scientists resulted in establishing the **International Society of Endocytobiology** and the interdisciplinary **International Journal of Endocytobiosis and Cell Research**. These institutions as a forum of communication may stimulate future research and thus contribute in achieving understanding of the problems of cellular biology such as cell differentiation (embryogenesis), cell **de**-differentiation (tumorigenesis) and cell rhythm (circadian rhythm).

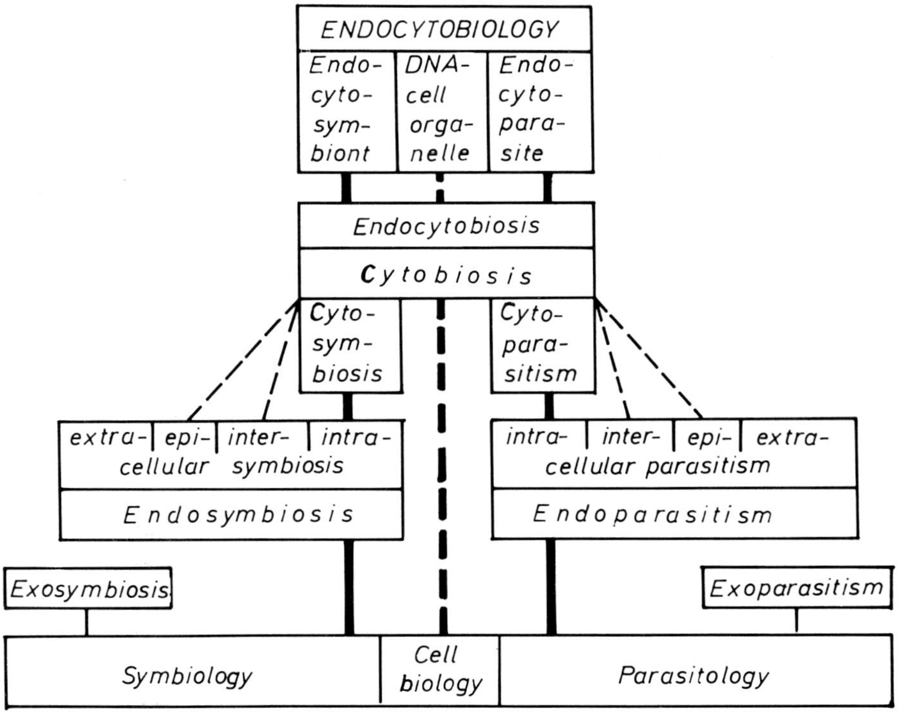

Fig. 80. Derivation and terminology of the new interdisciplinary research field, endocytobiology (Schwemmler et al., 1984).

IV. References

Altmann, R. (1890): *Die Elementarorganismen und ihre Beziehung zu den Zellen.* Veit and Comp. Leipzig.

De Bary, A. (1879): *Die Erscheinung der Symbiose.* Karl J. Trübner Verlag Straßburg.

Buchner, P. (1965): *Endosymbiosis of Animals with Plant Microorganisms.* Wiley and Sons New York.

Kleinig, H. and Sitte, P. (1984): *Zellbiologie.* Gustav Fischer Verlag Stuttgart.

Kraepelin, G. (1983): *Bypass respiration: Biological significance and evolutionary implications.* In: Schenk, H. and Schwemmler, W. (eds.): Endocytobiology II. Intracellular space as oligogenetic ecosystem, 291-302. Walter de Gruyter Verlag Berlin, New York.

Lee, J. and Fredrick, J.F., eds. (1987): *Endocytobiology III.* Ann. New York Acad. Sci. Vol. 503.

Margulis, L. (1981): *Symbiosis in Cell Evolution.* W.H.Freeman San Francisco.

Mereschkowsky, C. (1910): *Theorie der zwei Plasmaarten als Grundlage der Symbiogenesis.* Biol. CB 30, 278,321,353.

Nass, S. and Nass, M.M.K. (1963): *Intramitochondrial fibers with DNA characteristics, 2. fixation and electron staining reactions.* J. Cell Biol. 19 593-611.

Ris, H. (1961): *Ultrastructure and molecular organization of genetic systems.* Can. J. Gen. Cytol. 3 95-120.

Schenk, H. and Schwemmler, W., eds. (1983): *Endocytobiology II. Intracellular space as oligogenetic ecosystem.* Walter de Gruyter Verlag Berlin, New York.

Schimper, A.F.W. (1883): *Über die Entwicklung der Chlorophyllkörner und Farbkörper.* Bot. Z. 41 105-114.

Schwemmler, W. (1973): *In vitro Vermehrung intrazellulärer Zikadensymbionten und Reinfektion asymbiontischer Mycetocytenkulturen.* Cytobios 8 63-72.

- (1980a): *Modell zur molekularen Analyse von Circadianrhythmik, Einmusterbildung und Krebs.* Naturwiss. Rdschau 33 (2) 52-59.
- (1980b): *Endocytobiosis: General principles.* Biosystems 12 111-122.
- (1983): *Analysis of possible gene transfer between an insect host and its bacteria-like endocytobionts.* Int. Rev. Cytol. 14 247-266.
- (1984): *Reconstruction of cell evolution: A periodic system.* CRC Press Boca Raton, Florida.
- (1987): *Endocytobionts and mitochondria as determinants of leafhopper egg-cell polarity.* In: Frederick, J.F. and Lee, J., eds.: Endocytobiology III. Ann. New York Acad. Sci. 503 496-514.

Schwemmler, W., Hobom, E. and Egel-Mitani, M. (1975): *Isolation and characterization of leafhopper endosymbiont DNA*. Cytobiologie 10(2) 249-259.

Schwemmler, W, and Kemner, G. (1983): *Fine structural analysis of the egg cell of* Euscelidius *sp. (Homoptera, Cicadina)*. Cytobios 37 7-20.

Schwemmler, W. and Müller, H. (1986): *Role of insect lysozymes in endocytobiosis and immunity of leafhoppers*. In: Gupta, A.P., ed.: Hemocytic and humoral immunity in arthropods.chap. 17 449-460, John Wiley and Sons New York.

Schwemmler, W. and Schenk, H., eds. (1980): *Endocytobiology I. Endosymbiosis and cell biology. A synthesis of recent research*. Walter de Gruyter Verlag Berlin, New York.

Schwemmler, W., Schenk, H. and Smith, D.C., eds. (1984): *Introductory statement*. In: Endocyt. C. Res. 1 III-VI.

Taylor, F.J.R. (1974): *Implications and extensions of the serial endosymbiosis theory of the origin of eukaryotes*. Taxon 23 229-258

\- (1980): *The stimulation of cell research by endosymbiotic hypotheses for the origin of eukaryotes*. In: Schwemmler, W. and Schenk, H., eds.: Endocytobiology I. Endosymbiosis and cell biology. A synthesis of recent research, 917-942. Walter de Gruyter Verlag Berlin, New York.

Weber, H. (1966): *Grundriß der Insektenkunde*. Gustav Fischer Verlag Stuttgart.

Zabel, U. and Schwemmler, W. (1980): *Das Experiment: Insektenembryogenese am Beispiel einer Kleinzikade*. Biuz 10(4) 120-123.

Appendix

I. Author Index

II. Subject Index

The Author

Werner Schwemmler, born in 1940 in Germany, studied biology, chemistry and geography at the Universities of Marburg, Giessen, and Freiburg. He held stipendia from the "Duisberg-Stiftung" and the "Max-Planck-Gesellschaft" to work with Prof. Vago at the Institute for Comparative Pathology and at the Institute for Invertebrate Pathology of Montpellier University in France. The author took his doctorate in 1972 with Prof. Sitte, Freiburg.

From 1972 to 1974 he held a research fellowship for "Habilitation" from the Deutsche Forschungsgemeinschaft (DFG) at the Microbiological Institute of Prof. Drews, Freiburg. He spent some time at the University of Minnesota (Minneapolis) with Prof. Halberg and Prof. Brooks.

From 1974 to 1980 Dr. Schwemmler was an assistant professor at the Freie Universität of Berlin. He took his "Habilitation" in zoology there in 1975 and is the head of an interdisciplinary group working on the physiology, ecology, genetics and evolution of cellular systems.

He was co-initiator of the interdisciplinary approach in the new research area "Endocytobiology".
Since 1984 he has participated in the research project on "Structure and Evolution of the Cell" at the Freie Universität Berlin.